Elephant Trails

Animals, History, Culture

HARRIET RITVO, SERIES EDITOR

ELEPHANT TRAILS

A History of Animals and Cultures

Nigel Rothfels

Johns Hopkins University Press
Baltimore

 Published 2021
Printed in the United States of America on acid-free paper
2 4 6 8 9 7 5 3 1

Johns Hopkins University Press
2715 North Charles Street
Baltimore, Maryland 21218-4363
www.press.jhu.edu

Library of Congress Cataloging-in-Publication Data

Names: Rothfels, Nigel, author.
Title: Elephant trails : a history of animals and cultures / Nigel Rothfels.
Description: Baltimore : Johns Hopkins University Press, 2021. | Series: Animals, history, culture | Includes bibliographical references and index.
Identifiers: LCCN 2021006375 | ISBN 9781421442594 (hardcover) | ISBN 9781421442600 (ebook)
Subjects: LCSH: Captive elephants—History. | Elephants—History. | Human-animal relationships—History.
Classification: LCC SF408.6.E44 R68 2022 | DDC 636.96709—dc23
LC record available at https://lccn.loc.gov/2021006375

A catalog record for this book is available from the British Library.

Special discounts are available for bulk purchases of this book. For more information, please contact Special Sales at 410-516-6936 or specialsales@jh.edu.

In memory of Gunda, Alice, Josephine, Ned, Mena, Packy, Lily, and many more with and without names

CONTENTS

ACKNOWLEDGMENTS

I am grateful for support of this project from the US National Endowment for the Humanities and the Humanities Research Centre at the Australian National University in Canberra. I also received support from the University of Wisconsin–Milwaukee and want to thank, in particular, Provost Johannes Britz, Vice Provost for Research Mark Harris, and Deans Richard Meadows and Rodney Swain for their encouragement over many years. Staff at a wide range of archives, museums, and zoos have been very helpful, and I want to thank, in particular, Madeleine Thompson of the Wildlife Conservation Society at the Bronx Zoo, Mark Alvey of the Field Museum in Chicago, Peter Shrake of the Circus World Museum in Baraboo, Wisconsin, Mike Keele and Bob Lee of the Oregon Zoo, Richard Sabin and Laura McCoy at the Natural History Museum in London, Klaus Gille at Hagenbeck's Tierpark in Stellingen, Julie Irick at the Seattle Municipal Archives, and Darrin Lunde at the National Natural History Museum in Washington, DC. Any views, findings, conclusions, or recommendations expressed in this book do not necessarily represent those of the National Endowment for the Humanities or other organizations that have supported this work.

Elephant Trails

Blind Men's Elephants

There is an old story about religions, arguments, wisdom, truths, tolerance, and elephants. It is included among the scriptural works that pass on the teachings of Gautama Buddha.[1] In the story, followers of the Buddha tell him that in the local village, wanderers of various sects were arguing about religious truths. The Buddha responds to this information by telling a story of a king who had once gathered all the blind men in a village together and had an elephant brought before them. Each blind man put his hands on a different part of the elephant—a side, a tusk, the tail, a leg, the trunk, an ear. The king then asked each of the men to describe the elephant. Because each had touched only part of the great animal, each knew only a part of the truth. Soon fights broke out among the blind men because each described the elephant so differently. The Buddha explained that the wanderers of different sects were like the blind men: each was convinced that his partial knowledge was the whole truth, and so the squabbles began.

The poet John Godfrey Saxe used this story in "The Blind Men and the Elephant," his most popular work, subtitled "A Hindoo Fable." The poem begins:

> It was six men of Indostan
>   To learning much inclined,
> Who went to see the Elephant
>   (Though all of them were blind),
> That each by observation
>   Might satisfy his mind.[2]

The first man touched one side of the elephant and declared the animal was like a wall; the second felt a tusk and insisted the elephant was like a spear; the third put his hand on the trunk and was certain the elephant was like a snake; the fourth felt one of the elephant's legs and claimed it was like a tree; the fifth moved his hand across an ear and argued that the elephant was like a fan; and the sixth contended the elephant was like a rope after he touched the animal's tail. Saxe concluded:

And so these men of Indostan
Disputed loud and long,
Each in his own opinion
Exceeding stiff and strong,
Though each was partly in the right,
And all were in the wrong!

So, oft in theologic wars
The disputants, I ween,
Rail on in utter ignorance
Of what each other mean,
And prate about an Elephant
Not one of them has seen![3]

The story of the blind men and the elephant is fundamentally about the limits of human knowledge, about what is possible for us to know. The elephant stands for something beyond normal human comprehension, and over the last two thousand years the parable has usually been told as a lesson in humility and religious tolerance.[4] Although the story is not at its heart about knowing elephants, I think it might in a way be about knowing elephants, or, more broadly, about knowing across species. In 1922, Helen Keller visited the Bronx Zoo in New York with her three young nieces. There, in a moment likely organized by Elwin Sanborn (the zoo's official photographer), the deaf and blind woman who had become famous around the world for her lectures and writings reached out and touched the trunk of an elephant (fig. I.1). The "climax of happiness," she later recalled in an article for the *Bulletin* of the New York Zoological Society, came when she and her nieces "climbed up on the massive back of Alice, the kindliest of elephants."[5] That day at the zoo, Keller had touched a rattlesnake, was embraced by an orangutan, had fed giraffes—animals she described as "the saddest creatures under the sun"—and had even stroked the "wet furry coat" of the first living platypus to have ever been exhibited in the United States. And yet it was the moments with Alice, the elephant, that were particularly significant to Keller, to Sanborn, and probably to the readers of Keller's article, only some of whom would either have

Figure I.1. Touching an elephant. Photograph by Elwin Sanborn, © Wildlife Conservation Society, reproduced by courtesy of the Wildlife Conservation Society Archives.

been familiar with the account of the Buddha or Saxe's poem about the blind men and the elephant.

There was something about being close to an elephant that made Keller's moments with Alice the highlight of what was undoubtedly a trip filled with extraordinary experiences. She could not see Alice, the building where she lived, the keeper standing beside the animal, or the audience looking on; she could not hear the animal's rough skin rub against the howdah on which she and her nieces sat, the instructions of the keeper or Alice's rumbles and squeaks, the amazed comments of onlookers, or even the muffled sound of Alice's feet in the yard at the old elephant house at the zoo. While standing on the ground, Keller reached out and up and placed her left hand near the base of Alice's trunk. It was not a light touch; Keller reached confidently with her whole hand, and Alice looked down while her keeper, bullhook in his right hand, fed the elephant treats from his left. This was something Alice had done many times, but most visitors did not stand in front of her—she felt them as they climbed aboard the two-sided seat on her back—just visible at the top right of the photograph. Keller would have smelled the soft fragrance of hay so noticeable in the presence of captive elephants, a scent mixed perhaps with the odors of the dung and urine in the pen. She would have felt the warmth of Alice's rough skin, the power of her trunk, and the massiveness of Alice's skull as she reached from atop a platform to touch the top of the elephant's forehead before climbing aboard the howdah.

What was it about Keller's experience with Alice that made it more significant than being hugged in the arms of an orangutan or touching the hurried living body of a platypus, and how could Keller sense the "kindliness" of an elephant? I actually think Keller did sense something about Alice's personality that day during their brief encounter—felt her patience in standing calmly while being touched by a woman she had never met, intuited purposefulness in the elephant's movements as she carried Keller and her nieces, detected something, perhaps, in the vibrations of the rumbles coming from deep within the animal. I do not know if Alice was the kindliest of elephants. My guess is that she was less kindly than well trained from years of working as a ride elephant while she lived at the Bronx Zoo and at Coney Island's Luna Park before that. But Keller's truths, like those of the fable's blind men, do tell us part of the story of elephants and also part of the story of our thoughts about them.

Keller's encounter with an elephant points to an ancient fascination with the animal. From prehistoric cave drawings of mammoths in Europe and ancient art in Africa and India to burning pyres of confiscated tusks set aflame to protest the illegal ivory trade, we have used elephants to make sense of the world and our place in it. In *Les racines du ciel* (*The Roots of Heaven*), the French novelist Romain Gary's 1956 novel about World War II, colonialism, hope in a post-Holocaust world, and elephants, Gary made a simple point through his character Morel, a former French soldier, then POW in Germany, then fighter trying to save the last of Africa's elephants. Reflecting on times during which long-held beliefs in progress and the nation seemed shattered first by total war and then by unfulfilled promises after the war, Morel thought that people needed "something that can really stand up to it all." "Dogs aren't enough" to help people through their deep loneliness, he insisted. "What we need is elephants."[6] When Gary's novel was published, people apparently wondered what the elephants were meant to represent. In his introduction to the work's second edition, Gary addressed this question by explaining that he hoped the animals would be "a sort of Rorschach test"—that they would become what each reader needed. He concluded, "There is almost no limit to what you can make an elephant stand for."[7] In their great, gray opacity, in their distance from our usual familiars—dogs, cats, horses, and the rest—elephants seem to wait quietly for our interpretation, wait for Keller's touch, wait for us to make them into the monsters and miracles of our imaginations.

In this book, I trace a group of ideas about elephants—that they are wise and deeply emotional, that they have a special understanding of death, that they never forget, that they suffer unusually in captivity, and even that they are afraid of mice—to find out where "our elephants" come from. By "our elephants," I mean those elephants who actually live in the world today and also those who live only in our thoughts. These are the elephants described in nature documentaries, children's books, internet memes, *New Yorker* cartoons, safari tours, official campaigns in China designed to discourage people from purchasing ivory, policy discussions about the use of domestic elephants in Southeast Asia, arguments of protestors seeking to improve the conditions of elephants in captivity, and debates over the legalization of elephant hunting in areas in Africa. While ideas about elephants with deep roots in Africa and Asia have contributed to these accounts, I argue that most of today's globalized ideas about elephants can be traced through thousands of years of European history. Even though the ivory trade may now be driven by Asian markets, for centuries the lives of elephants hung more on Western actions and thoughts than on Eastern ones.

Elephants stand today at the center of a number of debates about the importance of conservation in a world of increasingly limited resources, about our responsibilities to animals in captivity and in the wild, about life and extinction. This book digs for the roots of contemporary ideas about elephants while insisting that our thoughts about animals are always historical. By history, I do not simply mean something of the past but also something that is embedded in our pasts, presents, futures, and cultures. Our ideas about anything, including elephants, are influenced by the historical circumstances of our lives, and most ideas about elephants today developed as part of worldwide changes that began in Europe in the late eighteenth century and continued through the early twentieth. This was a period that marked the beginning of what is usually described as the "modern" and that was distinguished by rapid urbanization, industrialization, militarization, and the expansion of empires; changes in family structures, education, recreation, and work patterns; reorientations in scientific theories and religious beliefs; and basic changes in life expectancy, foodways, and relations between genders, classes, and generations. Of course, thoughts about elephants did not remain unchanged over the thousands of years before modern times, and we continue to learn more about these remarkable creatures every day. Still, between the end of the 1700s and the first decades of the 1900s, thoughts about so much, including elephants, underwent profound changes. These changes left us the planet we inhabit today and these are the changes that have put the lives of elephants around the world in a persistent state of insecurity.

We seem to "know" about beings like elephants before we learn what science tells us about them. This knowledge flows from constellations of ideas that come down to us through time and culture—through stories we heard as children, through literally ancient accounts of them, through seeing them in captive settings or, in some parts of the world, in their native ranges, through attending to imaginative works about the animals, or through a kind of common sense that a creature like that just must think, feel, act, or be in certain ways. It is because our thoughts about elephants always have contours traced by our individual and collective histories that this is both a book about actual elephants who have lived or are living in the world and a book about what we believe about elephants. This book explores the origins of core contemporary ideas about animals whose natural and unnatural habitats include places like forests, savannas, circuses, zoos, works of fiction, hunters' memoirs, and the corners of human minds.

Accepting that there might be significant differences between our thoughts about elephants and their real lives is important if we want to look critically at what we think we know about them. While working on this book, for example, I have frequently been told in casual conversations that the amazing thing about elephants is that they mourn their dead. This is information that most people seem to think of as a discovery of recent decades, even though stories of elephant death rituals stretch back thousands of years. Moreover, not only is this idea not new but it is also unsupported empirically—the scientific basis for elephant death rituals is surprisingly thin. I hope through this book to give people a fuller sense of how these kinds of thoughts are part of history and to make a compelling case that a vital task before us today is to disentangle our own thoughts and wishes about elephant lives from the circumstances and challenges they truly face. In the chapters that follow, I examine key ideas that have long or more recently been associated with elephants. Although there are moments when I reach further back in time, the book is anchored in the critical period of European colonial expansion in Africa and Asia, especially the two hundred years before the beginning of World War I in 1914. The first two chapters trace ideas about elephants from classical times up to the second half of the nineteenth century—that they have a deep understanding of death, that they live in idyllic societies in closely knit families, that they are just and good, that they are spectacularly powerful but also benevolent, that they suffer—to set the background for the chapters that follow. In the third through sixth chapters, I dig into the origins of more recent ideas about elephants first by exploring accounts of the animals in classic memoirs of big-game hunters, then by presenting longer biographies of two elephants who lived most of their lives in zoos and circuses in the early decades of the twentieth century, and finally by considering the presence of elephants in early discussions of what we now call the sixth great extinction.

Toto

This project began for me when as a graduate student many years ago, I found a 1910 memoir titled *Wild und Wilde im Herzen Afrikas* (which translates as "wildlife and savages in the heart of Africa") by a man named Hans Schomburgk, a German hunter who later became a significant filmmaker and conservationist. In the book, Schomburgk relates that while hunting elephants for ivory, he became intrigued by an idea proposed by Ludwig Heck, then director of the Berlin Zoo, that someone should try to capture an elephant from a German colony to be exhibited in the German capital. After a number of failed attempts, in 1908 Schomburgk finally captured a young elephant in what was then the colony of German East Africa (now Tanzania) after having tracked him and his mother for three days. Coming across the two unexpectedly in high grass, Schomburgk quickly shot the mother head on. Another shot killed her, and, according to Schomburgk, the calf "remained standing beside its mother . . . hitting her with his trunk as if he wanted to wake her and make their escape."[8] Schomburgk included a photograph of the scene with the caption "Obeying the laws of nature, the young animal remained standing beside its mother."

Several months after capturing the young elephant, Schomburgk began walking him to the coast, and they arrived in Dar es Salaam in the fall. Despite his having suggested the idea of bringing an elephant to Germany, Heck did not respond when the hunter wrote to him to offer him the little elephant, and the animal was sold instead to an agent of Carl Hagenbeck, an animal dealer in Hamburg, Germany. In April 1910, Schomburgk visited Hamburg and saw the elephant again at Hagenbeck's revolutionary new zoo, where cages and bars were largely eliminated and animals were exhibited in naturalistic panoramas.[9] In November of that year, Hagenbeck sold the still young elephant along with two female elephants, Greti and Minnie, to the Rome Zoo, where they were on exhibit at the official opening a few months later. A photograph of Toto, as he became known in Rome, from soon after his arrival, shows him in fairly good condition in an outside yard, with a keeper on his back and in the company of Greti (fig. I.2).[10]

The early years of Toto's life in Rome seem to have been fairly incident free. According to the taxonomist and historian Spartaco Gippoliti, Toto went for walks around the zoo, interacted with the public and the other elephants, and was even on stage for the famous triumphal march in a performance of *Aida*.[11] Soon, though, his life began to change. In 1921 he killed a veterinarian treating

Figure I.2. Toto and Greti with keeper Angelo Pozzi in 1910–11. Courtesy of Spartaco Gippoliti.

an abscess on his shoulder, after which it was decided that he should no longer be taken out of his enclosure for walks around the gardens. Various keepers struggled to manage him, but eventually he seemed to respond to a keeper named Ivo Calavalle. Meanwhile, Greti died during World War I, and Minnie also died, reportedly as a result of aggression from Toto, in 1924. They were replaced by two male Asian elephants, Pluto and Romeo, but both of them died in July 1927. A female elephant was then apparently sent from England to keep Toto company, but she died in transit. Another female Asian elephant, Giulietta, arrived in the fall of 1927. By the time the Rome Zoo had its seventeenth anniversary, seven elephants had been acquired and only two were still alive. Twelve years later, on July 12, 1939, Toto died. He was about thirty-four years old.[12]

There have been many times over the years when I have thought about that photograph of Toto beside his dead mother. It is undoubtedly an extraordinary picture. I had already been researching and writing about the exotic animal trade in the decades before World War I for several years when I found it. I had seen plenty of photographs of animals after they had been brought back to the catcher's camp or arrived in Europe, but this photograph purports to show a young animal in the moments after his mother had been killed and before he was actually captured.[13] It is an image of extreme trauma, and many times over the years I have wondered about the ethics of using a picture like that in my work. This is a concern that was on my mind as I was writing this book as well. There are images and stories in this book that have challenged me because I am committed to exploring historic encounters between human and animal cultures but I also want to avoid just piling more abuse on top of historical injustices. I hope that through the histories I recount here I can expand on ideas regarding the importance and meaning of nonhuman lives in our past, present, and future without exploiting those lives for either gratuitous effect or to advance a personal, political, or ethical position.

When I found that first image of Toto now many years ago, I wondered whether it might be possible to write some sort of biography of him. As I began contemplating this book, I realized that the story I wanted to tell of this little elephant did not begin with the gun of a hunter in 1908 and it did not end with the animal's death decades later. When people get their pictures taken before a mounted mammoth skeleton in a museum, watch an elephant performance in a circus, book a weekend elephant adventure in Thailand or a safari in Africa, or point at an elephant in a zoo, they are all telling part of the story of Schomburgk's little elephant, Rome's Toto, and Keller's Alice, and all of these are blind men's elephants—creatures partially understood, described in ways that seem inevitably to tell us more about the observer than the observed. But if each of the blind men's perspectives is limited, I hope that by bringing their many viewpoints and experiences together, it may be possible to tell a richer collective story about the importance of these animals in our lives.

CHAPTER ONE

First among Monsters

In the back corner of a south London warehouse, a half dozen crates have been lined up against a wall. Each container has been placed on a pallet to keep it off the concrete floor and to permit ease of moving. The crates are different sizes and ages and variously made of black painted metal and dark woods that stand out against clean white walls. The last crate in the line, made of wood, is wider than it is tall. A hand-written label is affixed in the middle of its front panel with two tacks. In the center of the label are the words "ELEPHAS MAXIMUS," and at the bottom left are two simple words in a small neat script: "No history." At the top right is an accession number, and at the bottom right, it reads "Skull & skeleton." The lettering remains as crisp as the day the notations were written. On the crate itself, one can make out more letters in faded white chalk—in places barely visible and partly obscured by the label. The letters echo the label (fig. 1.1). *Elephas maximus* is the scientific name given by Carl Linnaeus in 1758 to what we now commonly call the Asian elephant. Plainly enough, then, the chest holds the skull and skeleton of an elephant. But what do the words "no history" hold?

Storage

The unmarked warehouse belongs to the Natural History Museum of London, an institution with one of the largest botanical, mineralogical, paleontological, and zoological collections in the world. Not surprisingly, most of the more than twenty-eight million zoological specimens in the museum's collection are insects and other invertebrates, but there are vast numbers of larger animals as well. A significant number of these specimens are cared for in this dry storage facility in the London borough of Wandsworth. In large rooms where

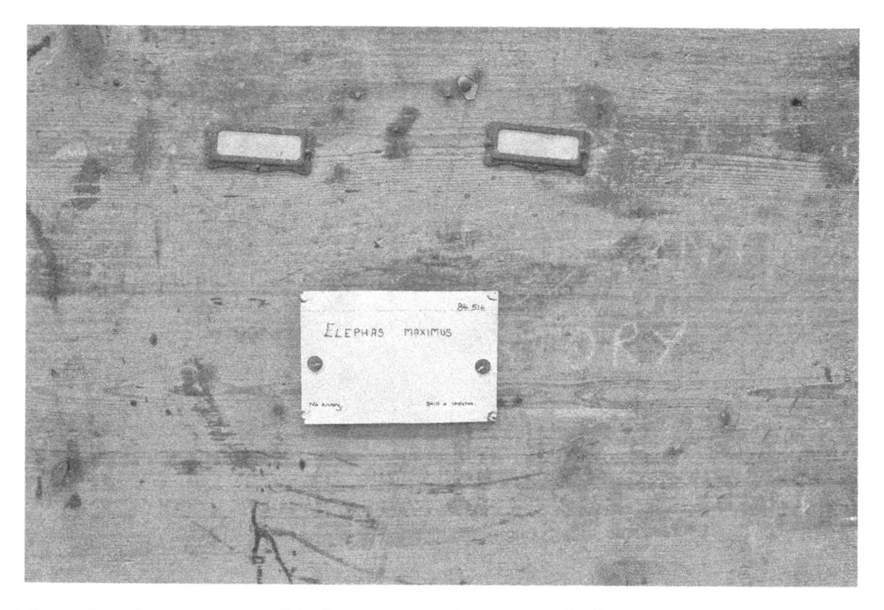

Figure 1.1. Storage crate with elephant remains, Natural History Museum, London. Photograph by Helen J. Bullard, with permission of the Natural History Museum, London.

temperature, humidity, and light are controlled, thousands of trophy heads, older taxidermy, bones, and articulated skeletons are "preserved . . . to all posterity."[1] The building is not open to the public, but it is a growing research collection, regularly accessed by scientists associated with or visiting the museum. One large room houses the whale collection; in another area, stuffed bears, once common in pubs, stand with iron armatures jutting out of their chests—supports for long-gone trays that used to hold glasses and whatnot; in yet another, one finds a collection of nineteenth-century stuffed domestic dogs representing the ideal conformations of major breeds. Elsewhere, a bank of taxidermized giraffe heads and necks rises like a formation of giant tubeworms surrounding a deep-sea vent. There are seals, zebras and other equids, all kinds of apes, the different species of pangolins, the world's bovines, and endless racks of trophy antlers. Helen and I are visiting to see and photograph the collection of elephant bones. Neither the tusk collection nor the elephant "spirit collection"—soft tissues preserved in formaldehyde, alcohol, and other liquids—are stored in these rooms. Here, though, on hundreds of linear feet of shelves, in cabinets, boxes, and crates, are elephant bones: rows of giant skulls, scores of mandibles lined up one next to another, a handful of complete and huge articulated skeletons, a young taxidermized elephant, and all manner of boxes and crates with jumbled up skeletons and parts of skeletons.

Even if you think you know what to expect from this place, it will take your breath away. At one level, it is utterly banal. Unlike the main museum in London with its grand Romanesque architecture, this warehouse is unmarked and unnoticeable, one of many such warehouses in the suburbs. We are buzzed through a security door and checked against a list of expected guests. An associate arrives to take us to the collection. We follow our guide to a set of large doors that open to the enormous rooms housing the dry remains of animals. The place is quiet and serious. Even while some of the items in the collection were specifically designed to be amusing, there is nothing light or frivolous about this place. We walk down long rows of shelving trying to get an overall sense of the place before turning our attention to the elephants. There is a weight here, a heaviness that we will try but largely fail to discuss with each other later; for now, though, we focus on the unusual opportunity before us and set up our equipment as we carry on a light conversation with the thoughtful young scientist who has been charged with helping us while making certain we abide by the rules. We know we are not typical visitors, but the fact that the collection has been made available to us points to its basic purpose—it exists to further knowledge. In our case the knowledge will not come from the analysis of a hair, bone, or skin sample. But what are we trying to know and how will we find the answers in this place? These questions linger at the edges of our conversations as we begin to take pictures of the remains of elephants.

Later that night, in an Indian restaurant in central London, Helen and I talk about our experience at the warehouse and their finding and photographing that crate that held "no history." We had each visited many museums of natural history over the years. We had each visited famous hunting collections, too, and had both written about taxidermy—we had come to this collection together because of our shared interests in these sorts of places and a wish to collaborate. We realized there was something about *this* place that was different, something about the quiet of it, something about the bright sterility of it, something about its scale and the staggeringly large number of parts of bodies in its rooms. Looking back, I think our reactions came from a mix of unease at the evidence of so much death and a genuine respect for the people and the institution committed to caring for these remains. The truth is that there were animal parts in this collection that had little apparent scientific value.

Gifts from this or that collector, the objects were here and not rotting in a landfill because the curators believed in their potential significance and, as importantly, were dedicated to the mission of stewardship.

Not surprisingly, this building made us think of other places, including cemeteries and ossuaries; it made us wonder, too, about the remains of the thousands of humans to be found in natural history museums and how our thoughts about those remnants both echoed and were nevertheless different from our thoughts about these elephant bones. As we gently opened a small nineteenth-century box containing bones from the skull of an elephant fetus, we struggled to contain questions about the lives intersecting in this improbable object. All by itself, this little box provides a unique intellectual, cultural, and material history. Most obviously, it is a lesson in changes in taxonomical nomenclature: the older label on the outside of the box provides the now anachronistic description "*Elephas africanus*"; the label on the inside, however, is updated to the current name: "*Loxodonta africana*." But there is more here. There is the box itself—the wood used to make it and how it was constructed; there is the name on the outside, Dr. Kirk; there is a history of a gift, an accession, a specimen's use by researchers for over a century. In addition, what does the fact that the little box is important at all to Helen and me say about *our* historical position? To put it simply, what is it that makes this little box different from any other box in this or any other warehouse?

In the end, it is the historicalness of this collection that fascinates us. There are parts of elephants here who were exhibited at the Zoological Gardens of London in the nineteenth century. There are the many skulls that hunters brought back as mementos of adventures—mementos that were turned over to the museum when changing ideas of decor rendered them obsolete as ornamentation. There are specimens that stand out for their size or peculiarity. Each has a history or rather histories, attested to by the tags, which recall morgue toe tags. These are not just any elephants; these are the remains of particular animals who, one way or another, have become part of lasting memory. But if that is the case, what does "no history" mean? When we opened the crate at the end of the line, we found a jumble of bones, still partly held together by desiccating and tightening connective tissues (fig. 1.2). These are not the carefully cleaned bones found elsewhere on the shelves; this crate smells of slow decay. Here, the words "no history" are used transparently, without a double meaning, to indicate only that the approximate date and location of the original specimen from which these bones were taken is not known to the museum. "History," here, is used in a way similar to that in the phrase "medical case history"—it refers to what is known of this specimen's origin. As far as the tag is concerned, what the crate contains is what it appears to contain: the jumbled-up bones of a smallish Asian elephant. One could know more, of course. One could examine the teeth and estimate the age of the elephant when it died; one could study key bones to determine whether the elephant was male or female; one could trace the accession number and discover when the specimen arrived at the museum and how; one could find out how the elephant has been used since it entered the collection; one could explore why the elephant's bones were boxed up to begin with and consider why it is important to keep this crate in a warehouse in Wandsworth. And beyond all this, beyond the circumstances that led to *this* elephant's bones being in *this* collection, the elephant had something like a prehistory—its life before it came into contact with humans.

This crate makes clear that in order to understand the elephants in *our* world (not just those in *the* world), we have to try to see history even when it seems it isn't there, even when it seems that all we are looking at is a very large animal or its jumbled remains. We must try to understand the significance of elephants in our thoughts and try to understand that these thoughts are present in historical

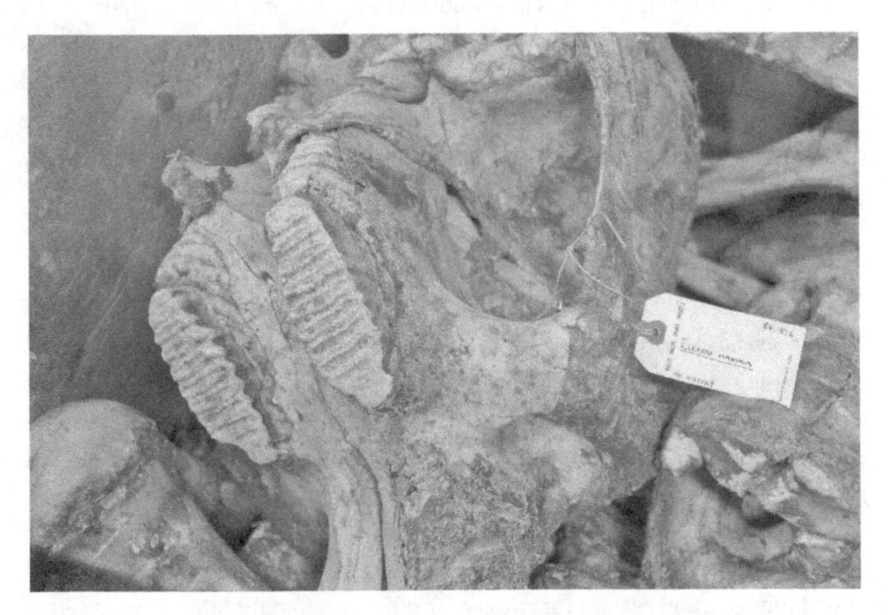

Figure 1.2. Elephant bones with no history, Natural History Museum, London. Photograph by Helen J. Bullard, with permission of the Natural History Museum, London.

contexts. To not notice the contexts is to mistake what we imagine an elephant is for what an elephant actually is. When it comes to the modern world and elephants, "no history" is simply an impossibility, a misapprehension that can have significant repercussions in the lives of actual elephants living in the world. It is not just that the Anthropocene makes it impossible for anything in the world to exist outside of human history; it is also that to a significant extent, we make the elephant world, if not the elephants themselves.

The Place Where Elephants Come to Die

> This marvel also I have heard, that the mighty elephants have a prophetic soul within their breasts and know in their hearts when their inevitable doom is at hand.
>
> —Oppian of Apamea, *Cynegetica*

The precise origins of the stories of Sindbad the Sailor, which describe events purported to have taken place during the late eighth century in the caliphate of Harun al-Rashid, will never be known. It is clear, though, that for centuries the stories were part of Arabic written and oral traditions. It is also clear that the stories first became widely known in the West at the beginning the eighteenth century when a French orientalist named Antoine Galland translated a manuscript version of the stories of Sindbad and included that translation as part of his twelve-volume *Les mille et une nuits* (1704–17, variously translated into English as *One Thousand and One Nights* and *Arabian Nights*). Over the course of the eighteenth century, Galland's work was translated into every major European language.

Many people today are at least passingly familiar with the stories of Sindbad and how he repeatedly succeeds in turning misfortune into wealth and fame. On his seventh voyage, Sindbad is traveling as an emissary of a sultan when marauders attack his ship. He is sold as a slave to a rich merchant who eventually puts Sindbad to work deep in the forest hunting elephants from a tree with a bow and arrows. Sindbad is remarkably successful and kills an elephant every day for two months. One day, however, the elephants do not just pass under his tree as usual; rather, they surround it, trumpeting loudly and staring at the horrified hunter. In the 1898 edition of the *Arabian Nights Entertainments*, compiled and edited by Andrew Lang, Sindbad explains: "I had indeed good reason for my terror when, an instant later, the largest of the animals wound his trunk round the stem of my tree, and with one mighty effort tore it up by the roots, bringing me to the ground entangled in its branches." Instead of being killed by the elephants, though, one of the elephants picks up the dazed Sindbad, and the herd carries him off into the forest. After that, Sindbad explains,

> it seemed to me a long time before I was once more set upon my feet by the elephant, and I stood as if in a dream watching the herd, which turned and trampled off in another direction, and were soon hidden in the dense underwood. Then, recovering myself, I looked about me, and found that I was standing upon the side of a great hill, strewn as far as I could see on either hand with bones and tusks of elephants. "This then must be the elephants' burying place," I said to myself, "and they must have brought me here that I might cease to persecute them, seeing that I want nothing but their tusks, and here lie more than I could carry away in a lifetime."[2]

Lang includes an illustration by Henry Justice Ford of the moment when Sindbad is lowered to the ground on the "Ivory Hill." Five elephants look down at the sailor, and in the background more can be seen to the horizon. Sindbad looks up at one of the elephants while resting on tusks beside a large skull; the animals regard Sindbad calmly, inquisitively, and benignly (fig. 1.3).

Galland's translation of Sindbad at the very beginning of the eighteenth century seems to have introduced the idea of elephant graveyards to the West, an idea that has had remarkable staying power. Zoologist Edmund Heller described the legend of the graveyards over two hundred years later in 1934 for *National Geographic*: "The tradition runs that when they feel death coming upon them, elephants leave the herd and trek to an elephant graveyard, a remote spot in the wilderness where all the elephants of the district go to die. There the ground is supposed to be thickly strewn with the huge bones of elephants, many having died 100 years ago or more." Heller makes clear that this is merely a tradition and insists that there is no record of such graveyards, despite the persistent efforts of ivory hunters to find them.[3] A piece published a year earlier in the *New York Times Magazine* entitled "Seeking the 'Ivory Valley,'" however, was less certain that the graveyards were only the stuff of legends. Pointing to the "facts" that frequent large caravans were observed carrying "old ivory" from the interior and that dead elephants were never found, the article claims confidently that elephants "know when death is upon them; and, in trumpeting the shrill call of death, they vanish into the secret valley where the huge skeletons of their forerunners lie whitening in the sun."[4]

As it turns out, Heller is not entirely correct that no one had ever discovered—or at least had claimed to discover—such a graveyard. From the nineteenth to

SEVENTH AND LAST VOYAGE 183

my terror when, an instant later, the largest of the animals wound his trunk round the stem of my tree, and with

SINDBAD LEFT BY THE ELEPHANTS IN THEIR BURIAL-PLACE

one mighty effort tore it up by the roots, bringing me to the ground entangled in its branches. I thought now

Figure 1.3. Henry Justice Ford, "Sindbad Left by the Elephants in Their Burial-Place," in Andrew Lang, ed., *The Arabian Nights Entertainments* (London: Longmans, Green, 1898). Special Collections, University of Wisconsin–Milwaukee Libraries.

the early twentieth century, the idea was regularly reinforced in accounts of hunters and adventurers. Usually, hunters had not seen these alleged burial grounds themselves and would instead recall stories they had heard. In his 1927 memoir of adventures, for example, Trader Horn mentions "elephant burial grounds," recalling that a man told him that old rogues would invariably hang around a "favourite ogey or spring of clear cool water generally in a grove," that wounded elephants "invariably died near the creek crossings or watering places" used for "bathing and cooling resorts," and that "old ivory, green and coloured ivory, was always dug up from around these places . . . and was always full-grown ivory."[5] Although most hunters reported stories they had heard, some claimed that they had actually found graveyards themselves. In his 1904 memoir *In Unknown Africa: A Narrative of Twenty Months Travel and Sport in Unknown Lands and among New Tribes*, Percy Horace Gordon Powell-Cotton describes discovering a plain covered with elephant bones. From the top of a rock mass that his guide called Ousereroc in British East Africa, Powell-Cotton surveyed the country below. He writes: "In all my journeyings through elephant country I do not think I had ever come across a skeleton of one of these beasts for whose death the guides could not account, and on no occasion did I see two skeletons together. Here I was surprised to find the whole countryside scattered with remains, the fitful sun, as it struggled through the clouds, lighting up glistening bones in every direction." The hunter adds that his guide "called this 'The place where the elephants come to die'" and assured him that "it was no fell disease which had decimated a vast herd," which is what the hunter had at first thought; rather, the guide insisted that "when the elephants felt sick, they would deliberately come long distances to lay their bones in this spot."[6] Powell-Cotton declares that before he had seen the proof "before my very eyes," he had regarded the stories of elephant graveyards as myths, but from that day forward he was convinced.

There is more to the story of elephant graveyards, however, than a reference in the tales of Sindbad or the exaggerations of swaggering hunters and explorers. At the heart of the story is a cluster of broader ideas about elephants and a spiritual realm that has persisted for thousands of years. A hint of those ideas can even be found in the ancient image of gates of ivory and horn in Homer's *Odyssey*. In book 19, Odysseus has returned in disguise to his wife, Penelope, to find her surrounded by suitors whom she has so far successfully kept at bay. Alone together, Penelope tells the "stranger" of a dream she had had about an eagle from the mountain coming to kill twenty geese in her house in which the eagle turned out to be her long-lost husband come to kill her suitors. Odysseus replies that the meaning of the dream seems clear enough. Penelope, however, retorts that the meaning of dreams can be deceptive: "Dreams," she says, "are baffling and unclear of meaning, and in no wise do they find fulfillment in all things for men. For two are the gates of shadowy dreams, and one is fashioned of horn and one of ivory. Those dreams that pass through the gate of sawn ivory deceive men, bringing words that find no fulfillment. But those that come forth through the gate of polished horn bring true issues to pass, when any mortal sees them."[7] As visions and promises, then, dreams can be true or false and enter the world through different gates of horn and ivory.

Over the almost three thousand years since Homer, the image of the gates of ivory and horn has been recalled by many writers. Perhaps most famously, at the end of the first half of Vergil's epic poem *Aeneid* (19 BCE), the hero Aeneas stands before the two gates. Aeneas has been visiting the underworld in the company of his deceased father, Anchises, and the Cumaean Sibyl and is ready to leave the world of shades:

> There are two gates of Sleep: the one is said
> to be of horn, through it an easy exit
> is given to true Shades; the other is made
> of polished ivory, perfect, glittering,
> but through that way the Spirits send false dreams
> into the world above. And here Anchises,
> when he is done with words, accompanies
> the Sibyl and his son together; and
> he sends them through the gate of ivory.[8]

Just why Vergil has Aeneas and the Sibyl leave the underworld through the gate of ivory—the gate of false dreams—has been debated for centuries.[9] But the question of why Homer (and then later Vergil) described the gates as being made of horn and ivory has received less attention. The leading argument seems to be that the gate of horn developed out of ancient Egyptian and Mesopotamian ideas about gates serving as portals for gods and spirits and the bull being deeply connected to mythic stories of life and death. In his study of the archaeological backdrop to the gates, for example, Ernest Highbarger concludes that Homer used the idea that the abode of the dead is "approached by a 'Gate of the Horns.'," which was "conspicuous in the Orient from earliest times" and "equally important in Crete and Greece."[10] Highbarger argues that the gate of ivory developed from earlier conceptions of a "gate of the sun"—which was a portal for the gods—and that ivory was chosen to adorn the gate because of its rarity and value and because its white brilliance suggested the "clouds" of the realm of the gods.

In ancient Greek, the word "eléphas" (ἐλέφας) translates as both "ivory" and "elephant." Ivory, which appears to have been known before there was a widespread understanding of elephants themselves, was not simply a part of the elephant; in a way it was the elephant itself.[11] In light of this, the choice of ivory for the gate makes sense because once the rare material was connected to elephants it also became connected to developing ideas about the relationship between elephants and spiritual practices. The key conduit here is Pliny the Elder (first century CE), who focused on the behavior of the animals. In his *Natural History*, Pliny writes that elephants "have been seen when exhausted by suffering (as even those vast frames are attacked by diseases) to lie on their backs and throw grass up to the heaven, as though deputing the earth to support their prayers."[12] Elephants, he argues, have a "religious respect also for the stars, and a veneration of the sun and the moon."[13] He continues: "at the first appearance of the new moon, herds of these animals come down from the forests of Mauritania to a river, the name of which is Amilo; and that they there purify themselves in solemn form by sprinkling their bodies with water; after which, having thus saluted the heavenly body, they return to the woods, carrying before them the young ones which are fatigued. They are supposed to have a notion, too, of the differences of religion."[14] Pliny also claims that elephants know that hunters seek them for their ivory. He notes that "these animals are well aware that the only spoil that we are anxious to procure of them is the part which forms their weapon of defence" and claims that "when their tusks have fallen off, either by accident or from old age, they bury them in the earth."[15]

A century later, Aelian similarly observes that when an elephant "sees another lying dead, it will not pass by without drawing up some earth with its trunk and casting it upon the corpse, as though it were performing some sacred and mysterious rite on behalf of their common nature; and that to fail in this duty is to incur a curse. It is enough for it even to cast a branch upon the body and with due respect paid to the common end of all things the Elephant goes on its way." Then, expanding on Pliny's account, he adds that "when Elephants are dying of wounds, stricken either in battle or in hunting, they pick up any grass they may find or some of the dust at their feet, and looking upwards to the heaven, cast some of these objects in that direction and wail and cry aloud in indignation in their own language, as though they were calling the gods to witness how unjustly and how wrongfully they are suffering."[16]

Even though, therefore, the concept of elephant graveyards did not establish a foothold in Western thought until the eighteenth century through the popularization of *One Thousand and One Nights*, many elements of the story are present in much older Western accounts. Both Pliny and Aelian reference already established beliefs that elephants have a unique comprehension of mortality, are aware of their circumstances when faced with illness or injury, engage in ritualistic or religious activities, understand that they are killed by humans for the sake of their ivory, are able to understand that the corpses of elephants are the remains of once living animals, treat the bodies of deceased elephants with special regard and attempt to cover them with earth, and even play a role in our own understanding of death and the afterlife. These ideas, it should be said, were not based on the empirical observations of these ancient authors; they were instead embedded in their cultures. When Aelian relates that elephants plead to the gods and pay respect to the mortal remains of dead elephants and when Pliny tells how elephants ritualistically bathe and worship celestial bodies, we catch glimpses into how ancient Romans sought to understand the lives and minds of elephants and how those efforts were always a part of understanding human lives and thoughts. It is *not* that the descriptions of elephants in these texts are fictional or were not ultimately based observations of actual elephants; no doubt, these descriptions were informed by real-life experiences with elephants, but these stories also tell us something significant about what the authors believed about the lives of elephants, just as they also tell us something about Roman culture—these are accounts of elephants, but they are much more than that.

The Elephant Corpse

If an eighteenth-century Sindbad and nineteenth-century hunters told stories of elephant graveyards, in the second half of the twentieth century, in a time of science and the fading of mystery, the idea of a secret valley for dying elephants as depicted in the 1930s *Tarzan* movies became increasingly less plausible. With that said, interest in how elephants experience and understand death did not diminish. In the classic and profoundly influential 1975 memoir *Among the Elephants*, written by Iain and Oria Douglas-Hamilton, for example, Iain Douglas-Hamilton explains how he began to wonder about elephant death. At one point, he had been checking in daily to record the decomposition of an elephant carcass. After ten days, he writes, the body was "reduced to a foul black cavity enclosed by a bag of skin, with bones sticking out."[17] On that morning, though, a large group of elephants led by the matriarch Clytemnestra came up the path near the corpse while he was watching. When she caught wind of the body, she quickly turned. Douglas-Hamilton writes, "Her trunk held out like a spear, her ears like two great shields, she strode purposefully toward the scent, like a medieval olfactory missile of very large proportions." Three other cows joined her. He continues:

> Their trunks sniffed at first cautiously, then with growing confidence played up and down the shrunken body, touching and feeling each bared fragment of bone. The tusks excited special interest. Pieces were picked up, twiddled and tossed aside. . . . Before this incident, I had heard of the elephant's graveyard, the place where elephants are supposed to go to die. This persistent myth I knew to be untrue after discovering elephant corpses scattered all over the Park. I had, of course, also heard that elephants took a special interest in the corpses of their own kind; it had sounded like a fairy tale and I had dismissed it from my mind. However, after seeing it with my own eyes I collected every reliable account I could find.[18]

After reviewing what others, including David Sheldrick in the 1950s, had reported, Douglas-Hamilton decided to perform a series of what he called "crude" experiments of putting elephant bones on the trails of elephants to observe how the animals might react. One of these experiments he organized as part of filming for a television documentary. In his characteristically evocative style, he describes the encounter of the also remarkably named Boadicea and her group with bones, and it is worth quoting at length:

> It seemed at first that they would pass the corpse. Then a breath of wind carried its smell directly into their trunks. They wheeled *en masse* and cautiously and deliberately closed in on the body. Shoulder to shoulder the front rank drew nearer, ten trunks waving up and down like angry black snakes, ears in that attentive half-forward position of concern. Each individual seemed reluctant to be the first to reach the bones. They all began their detailed olfactory examination. Some pieces were rocked gently to and fro with the forefeet. Others were knocked together with a wooden clonk. The tusks excited immediate interest; they were picked up, mouthed, and passed from elephant to elephant. One immature male lifted the heavy pelvis in his trunk and carried it for fifty yards before dropping it. Another stuffed two ribs into its mouth and revolved them slowly as if he were tasting the surface with his tongue. The skull was rolled over by one elephant after another. To begin with only the largest individuals could get near the skeleton, such was the crush. Boadicea[,] arriving late, pushed to the center, picked up one of the tusks, twiddled it for a minute or so, then carried it away, the blunt end in her mouth. The rest of the group now followed, many of them carrying pieces of the skeleton, which

> were all dropped within about a hundred yards. . . . It was an uncanny sight to see those elephants walking away carrying bones as if in some necromantic rite.[19]

The key points in this story—the elephants becoming aware of the corpse through smell; their cautious approach; their examination of the remains by smell, touch, and taste; their manipulation of the bones with their trunks, mouths, and feet; the particular fascination with the tusks; and the carrying away of bones and tusks—have been recorded time and again by observers in recent decades. As astonishing as his account is, though, Douglas-Hamilton is careful to retain his scientific distance on what we might call the "corpse question" and also reports another experiment where similar behaviors were seen in only six of eight groups of elephants that passed the remains. For Douglas-Hamilton, as important as the behaviors of the six groups were, he is also intrigued by the two groups of elephants who did not react to the bones, who "simply walked over the bones as if they were not there."[20]

Similarly, although Douglas-Hamilton is struck by the ubiquity of reported burying behaviors of elephants—when coming across a dead body, even that of a human, elephants will seek to cover the remains with dirt, mud, and plants—he is clear that despite years of observation he has never seen the behavior and he resists offering an explanation for it. His conclusions in this account seem to be constrained by, first, his commitment to a scientific method that requires compelling and replicable experiments and, second, an understanding of evolutionary theory that requires behaviors to be adaptive to the survival of the species. He does accept that despite the limited nature of the evidence "it is clear that elephants often show more than a passing interest in decomposed carcasses of their own species, even when little remains but the smell." As for the carrying of bones and ivory, he admits that he has "no idea" why elephants sometimes do this, although he thinks the "special significance of the tusks" might be explained by the fact that they "remain much the same in death as life, curving shafts of ivory, perhaps still with some signal effect."[21] In any event, he insists, "it is not enough to say that an elephant possesses a 'sense of death,' and to leave it at that."[22]

If for Douglas-Hamilton in the mid-1970s there were few scientific grounds for speculating about elephants' special "sense of death," researchers who have followed him have often been more amenable to that argument. In this, as in many other areas, the exceptional fieldwork conducted over four decades by Douglas-Hamilton's former assistant Cynthia Moss, cofounder in

1972 of the Amboseli Elephant Research Project at Amboseli National Park in Kenya, has been centrally important. In her 1988 account of her work, *Elephant Memories,* Moss once again gives a nod to the old legend: "Elephants may not have a graveyard but they seem to have some concept of death. It is probably the single strangest thing about them. Unlike other animals, elephants recognize one of their own carcasses or skeletons. Although they pay no attention to the remains of other species, they always react to the body of a dead elephant. I have been with elephant families many times when this has happened."[23] She writes that when elephant groups "come upon an elephant carcass they stop and become quiet and yet tense in a different way from anything I have seen in other situations. First they reach their trunks toward the body to smell it, and then they approach slowly and cautiously and begin to touch the bones, sometimes lifting them and turning them with their feet and trunks. They seem particularly interested in the head and tusks and lower jaw and feel in all the crevices and hollows in the skull. I would guess they are trying to recognize the individual."[24] Moss also describes elephants burying their dead with dirt and palm fronds and a particular occasion when she was certain that a seven-year-old elephant recognized the jawbone of his dead mother.[25]

In many ways, Moss echoes Douglas-Hamilton's observations, but she goes further in her conclusions. Convinced that elephants are more than just "interested" in the remains of other elephants, she argues that the animals are able to distinguish elephant remains from those of other animals, that they are only interested in the remains of elephants, that they consistently react to the bodies of dead elephants, and that they seem to be able to recognize individuals based on their remains. Moss has since moderated some of these claims. In the early 2000s, she, Karen McComb, and Lucy Baker, undertook an updated version of Douglas-Hamilton's experiment with free-ranging elephants in Amboseli. They presented elephants "with animal skulls, ivory and natural objects" to investigate whether they would be "attracted to elephant skulls and ivory over other objects," would show "more interest in elephant skulls than in skulls of other large terrestrial mammals," and would especially pick out "the skulls of relatives for investigation."[26] The results showed that when confronted with a tusk, an elephant skull, and a piece of wood, elephants were most interested in the tusk and least interested in the wood; when presented with the skulls of an elephant, a rhino, and a buffalo, the elephants were most interested in the elephant skull; and finally when presented with the skull of a deceased matriarch of their own kin group along with the skulls of two other matriarchs, the elephants did not exhibit any particular interest in one skull over the others—they did not, that is, obviously "distinguish" the skull of their family member. There are reasons to wonder about these results, however, as the study accounts for neither the novelty of the objects in the landscape nor the inherently different and probably more interesting nature of elephant remains, if only because of their size. What would happen, I wonder, if scientists put a whale's skull beside an elephant's skull on the elephant path. My guess is that the curious elephants would find the whale skull fascinating, probably worth exploring with their trunks, mouths, and feet, and perhaps even worth trying to carrying around for a while. But what could or should we conclude from that? Joyce Poole and Peter Granli have recently stepped back even further on the idea that elephants recognize deceased predecessors and have a unique understanding of death, but they nevertheless continue to cite the work of Douglas-Hamilton and McComb, Baker, and Moss, relying on it to argue, for example, that elephants "typically also investigate the remains of elephants sniffing with the trunk, using the feet and trunk to Explore-Touch, lifting, carrying, and playing with the bones or considering them in quiet reflection."[27]

The conclusions and informed intuitions of people like the Douglas-Hamiltons, Moss, and Poole, along with those of other field researchers discussed in this book, simply must count for something. It is obviously true that we often incorrectly interpret the intentions and behaviors of all sorts of animals around us (including those of our fellow humans). It is also true, though, that some of our guesses or conclusions, some of our leaps of empathy and imagination, are correct. Throughout the time I was conducting research for this book, I heard people begin sentences with the phrase, "Well, if I were an elephant . . ." Our efforts to imagine ourselves into the worlds of other animals, even if we recognize that their perceptual and experiential worlds can be so extremely different from ours, is not the worst part of anthropomorphism, as it merely suggests our desire to use empathy to understand them. Nevertheless, we must also grant that many of the questions we ask about elephants stem from matters that interest us and not necessarily elephants themselves. In the end, we ask questions about whether elephants have a unique understanding of death because questions of death are particularly important to *us*, and so when we think we see an animal expressing an interest in death, we become very attentive. I have often imagined that a researcher observing elephants in the wild carrying and chewing on a large branch from a tree for many hours probably simply documents in her log that the elephants continue to forage but then starts to record every gesture the moment an elephant turns to look at bleached bones on the path.

Moss suggests that there are perhaps foundations to the myth of elephant graveyards. She points out, for one, that hunters sometimes herd elephants together for slaughter, leaving skeletons concentrated in a certain area; for another, she notes that "in any given elephant range there are places the sick and wounded elephants tend to go. These would be areas where there is water and shade and soft vegetation to eat." When Amboseli elephants are too ill to travel, she explains, they often stay close to the shade of the acacia trees around the swamps near her camp. "Places such as this," she concludes, "might have more carcasses than other parts of a population's range, and therefore people may have thought there was a special area where elephants went to die."[28] At one level, Moss provides here the sort of sensible response to the eighteenth- and nineteenth-century accounts like those found in Galland's and Trader Horn's work that one might expect to find in the twentieth and twenty-first centuries; a long-standing mystery is resolved by a reasonable conclusion based on the observation that weak elephants tend to seek out and stay around water and easy access to food and shade. On another level, though, Moss's thoughts about elephants are part of a much older preoccupation with elephants and death, a preoccupation that dates back to before there were written texts in the West about the animals. I am not saying that Moss or Douglas-Hamilton are wrong when they suggest that elephants have an unusual fascination with the bodies of dead elephants; I am saying that the fascination with this issue is far more than a matter of pure scientific interest.

Monstrorum princeps

In the thirteenth century, Bartholomaeus Anglicus retold Pliny's story of elephants ritualistically gathering in the light of the moon: "Among beasts the elephant is most of virtue, so that unneth among men is so great readiness found. For in the new moon they come together in great companies, and bathe and wash them in a river, and lowte each to other, and turn so again to their own places, and they make the young go tofore in the turning again; and keep them busily and teach them to do in the same wise: and when they be sick, they gather good herbs, and ere they use the herbs they heave up the head, and look up toward heaven, and pray for help of God in a certain religion."[29] Bartholomew claims that elephants have more virtue than any other animal, and that even among men such a measure of virtue cannot be found. He writes about fantastic creatures, creatures who gather in sacred places in the moonlight, use medicinal herbs, and teach their young to pray. Just because he called his work a "natural history," though, does not make it the equivalent of the works of Douglas-Hamilton or Moss—his account served an entirely different purpose and was based on very different kinds of observations of the world. But if Bartholomew's work is not exactly science, as we understand that concept today—as a body of objective knowledge and methods—our current ideas about elephants and death likewise go well beyond science. Bartholomew's argument about elephants and morality, Aelian's description of elephants covering their dead with earth and branches, legends of elephant graveyards, and even the shelves and crates of elephant bones in elephant graveyards like the Natural History Museum in London all contribute to the idea that elephants are not simply taxonomically, physically, intellectually, or emotionally different from other creatures of the world but are fantastically unique.

I had a conversation once with the biologist J. Rudi Strickler, much of whose long career has been dedicated to studying copepods—the small freshwater and marine crustaceans that constitute a large part of the animal biomass of the earth.[30] When we started talking about elephants, Strickler chuckled and observed that the significance of elephants tends to be as exaggerated as their size. What he meant is that the presence of elephants on the planet is not as important (to all other life) as the presence of the very small organisms like copepods that make the life of everything else possible. Perhaps it is unfair to compare the impact of over 20,000 species of copepods with that of three species of mammals, but one is still forced to grant that if the three extant species of the taxonomic order Proboscidea went the way of the over 175 extinct species of the order, the planet might not be fundamentally changed. Strickler is obviously right that we tend to exaggerate the importance of elephants; he may also be right that we do so at least partly because of their "exaggerated" size, because they are the biggest land animal alive today and thus have an unusually charismatic presence in our imaginations.[31]

There is an old Roman epigram about elephants that describes the creatures as spectacularly dangerous but also as sources of pleasure when their ivory is turned to game pieces. At its end, the poem draws a conclusion about the ephemerality of power in human affairs:

> First among monsters, the elephant, fierce with trunk,
> bristles with black massive bulk, gleams with snowy tusk.
> But even though one must flee the brute
> blazing up in anger with unpredictable destruction,

the death of the caught wild animal is valuable.
For the tusks we see of mountainous strength
become well suited for human uses.
From them come the scepter for consuls, decorations for tables,
the weapons of the dice throwers, and the varicolored pieces of the game board.
This is the constant changeability of the human condition:
That which was once a source of fear becomes in death an amusement.[32]

Edward Topsell, the seventeenth-century English compiler of natural history, notes in his *History of Four-footed Beasts, Serpents, and Insects* that epigrams and scorpions are similar: just as "the sting of the Scorpion lyeth in the tayl, so the force and vertue of an Epigram is in the conclusion."[33] The lesson of the epigram is, indeed, to be found in its tail—in the final couplet's age-old lesson that the nature of human fortune is capricious, that those with fame and power today will be the objects of laughter tomorrow.

The phrase "arma tablistis" ("the weapons of the dice throwers") in the poem recalls Pliny's use of the word "arma" to describe the weapons of the elephants—their tusks. The weapons, then, of the "first among monsters," its ivory tusks, become the weapons of the dice throwers, their ivory dice. In becoming dice—and to this day "ivories" remains slang for "dice"—the power of the monster is thus rendered an amusement, a mockery. This idea is echoed in a 1929 article in the *New York Times* entitled "Where Do Dead Elephants Go?" In the piece Ernest Shaw offers his thoughts about the legendary elephant graveyards, relating that when he was living on the west coast of Africa he would ask "the natives" where elephants die. He reports that he always received the same answer: "Elephant does not die, we kill him." Over time, Shaw writes, he came to the conclusion that "elephants rarely die of old age, but fall prey to their only enemy, man, when they grow too old to protect themselves." And the reason one never finds traces of dead elephants, he argues, is that just "as the butcher prides himself on utilizing everything that goes to make a pig except the squeal, so it is with these natives. They eat the flesh and hide, sell the ivory or turn it into ornaments, make bracelets and rings out of the few hairs an elephant possesses, and let the children make toys with the smaller bones. Other bones make good chewing for the village dogs and those that defy the canine jaws will be greedily accepted by the ravenous jackals and hyenas."[34]

"We kill him," and his remains end up in our homes and museums as objects of science, contemplation, and amusement. Years after the visit to the Natural History Museum in London and in another off-site storage facility belonging to another museum of natural history—this time the Smithsonian Institution's National Museum of Natural History—I find a bone of an African elephant resting on white foam. I am thrown back to my earlier experience by a familiar tag: "No history" (fig. 1.4). Again, the tag contains more information than that. Printed at the top is the specimen's catalogue reference and the symbol for female; the scientific name ("LOXODONTA AFRICANA") appears below that; and at the bottom it reads "Philadelphia Zoological Gardens" and "died 12 Mar 1943." The bone is one of many in the collection from this elephant. Soon I find her other leg bones, her vertebrae, ribs, mandible, and then her white skull. Finally, at the back of a locked cabinet, I find her long, thin, arching tusks (fig. 1.5).

The date and the reference to the Philadelphia Zoological Gardens make it a simple enough task to learn more about this elephant's life. The remains come from an elephant named Josephine who came to the zoo from West Africa as a young elephant in 1925 to be exhibited in the children's zoo. Upon her death, an Associated Press article published in papers across the country memorialized "the toy elephant that grew up." According to the article, Josephine, who was publicized nationally as a "pygmy elephant," was the "prima donna of the baby pet zoo." Over the course of her eighteen years in Philadelphia, officials estimated, she had given rides to about 175,000 children. She was the "only forest elephant in the new world" ("one of a slender type found in sparsely forested parts of West Africa") and was the official mascot of the Republican National Convention in Philadelphia in 1940. When in that year an "ultra-modern cage" was built for her, the article reports, she apparently "didn't like it and began a three-week sit-down strike. Then, late one night, she went AWOL, knocked over stands and ate tree foliage," before returning "docilely into the cage." "Yesterday," the article concluded, "at 20 years of age, after a long illness of heart disease, she died."[35]

There are people alive today who might remember seeing Josephine at the zoo, who might remember riding her, too, even though they might not know she had a name, that she became a mascot for a national political convention, or that one day she got loose and caused a little havoc. They would not know, either, that her remains are housed today in "Pod 2" at the Smithsonian's Museum Support Center in Maryland and that those remains would be better labeled *Loxodonta cyclotis* (African forest elephant) than *Loxodonta africana*. The white, stained bones come from Josephine's body, but they are both less and more than her. Less than her, obviously, because so much of what made her who she was has been lost—even her identity as a particular elephant who

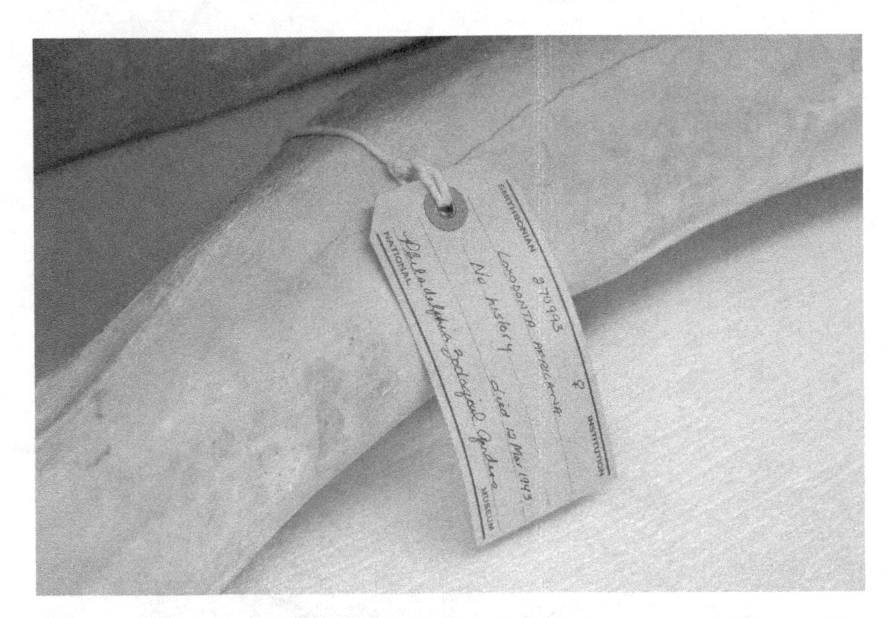

Figure 1.4. Bone with no history, National Museum of Natural History. Photograph by Nigel Rothfels.

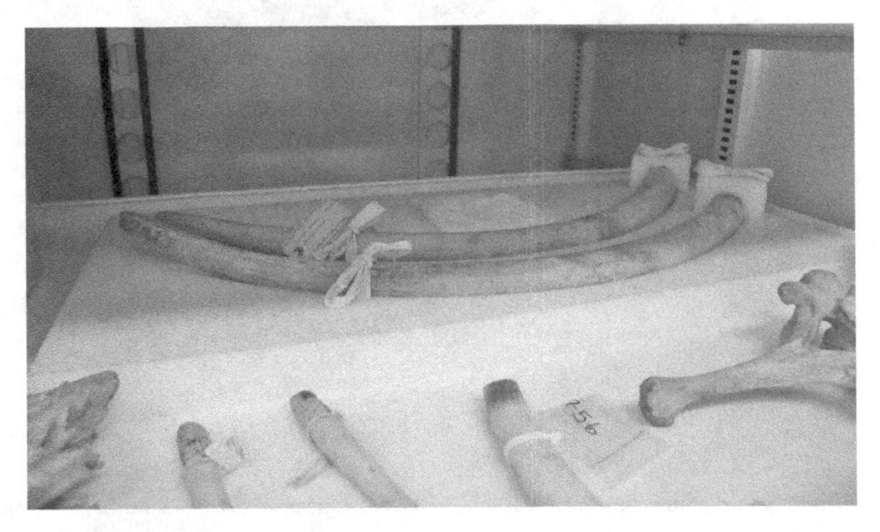

Figure 1.5. Josephine's tusks, National Museum of Natural History. Photograph by Nigel Rothfels.

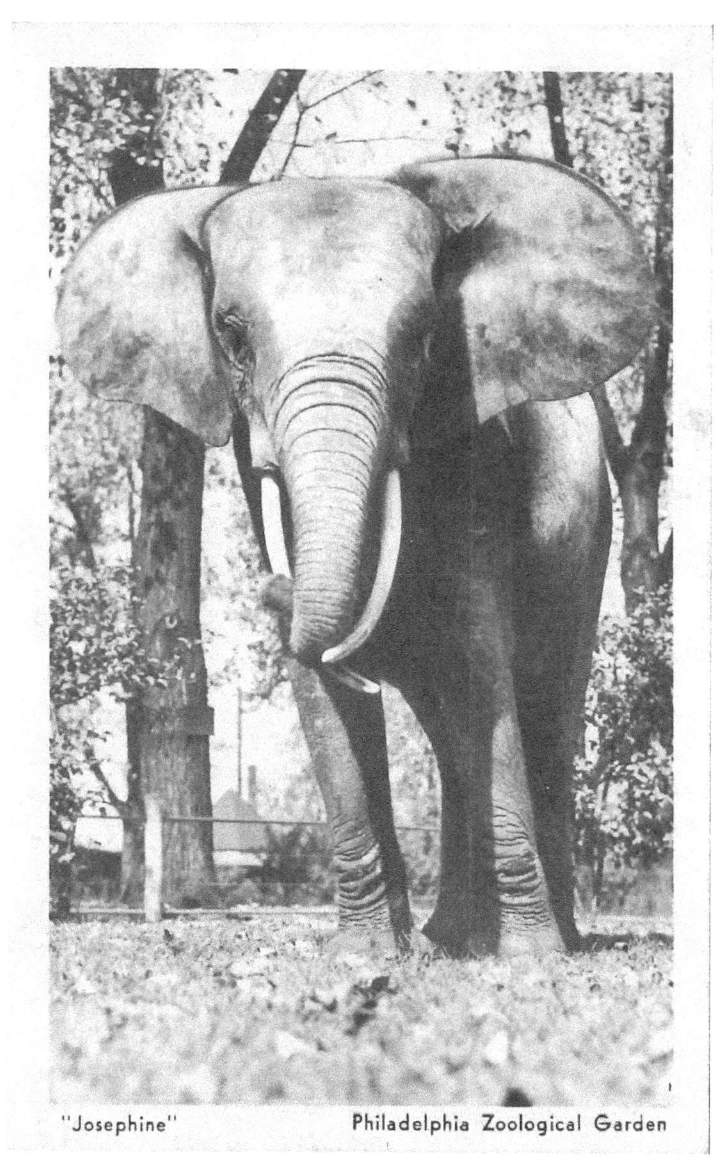

Figure 1.6. Josephine, Philadelphia Zoological Garden postcard, ca. 1943. Author's collection.

charmed the children of Philadelphia. But more, too, because her remains have become part of a very old and very deep history of ideas about elephants and death. We, of course, recognize the tag as a text, as something made possible by and part of human culture, as a human artifact, as something we are meant to interpret, but in many ways the bones are as well. Housed in a special kind of ossuary, Josephine's bones are part of human culture, preserved right along with a history of ideas. I eventually picked up a postcard of Josephine on eBay (fig. 1.6). The orientation is portrait, and she has been photographed head on, her round ears out. She appears relatively small standing on closely cut grass; behind her are trees and a low, railed fence. Her long tusks, the tusks I had seen in the museum, cross and she has twisted the end of her trunk, an action that, along with the position of her ears, suggests that she is feeling apprehensive.[36] It was probably unusual for her to be asked to stand still and alone like this before a photographer—normally one of the keepers would have been beside her, talking to her, reassuring her, instructing her, giving her a sense of clarity in a confusing moment. The printed text on the back of the card reads: "FOREST ELEPHANT Central Africa. Despite her large size and imposing appearance Josephine is a very gentle animal. Each day in seasonable weather she is saddled and taken to the riding rings where children and adults are permitted to ride on her." The card was postmarked July 10, 1945, more than two years after Josephine's death, and includes a simple message to Mrs. Lillie Stoudt of Reading, Pennsylvania, written in what appears a child's hand: "We are having a good time at the zoo. Nancy and Judy and Linda."

Is it simply a projection of emotion that leaves me with feelings of sadness when I see images like this one of Josephine? I cannot really know what she was feeling that day in front of the camera. If the position of her trunk and ears suggests apprehension, is that apprehension fearfulness? Worry? Concern? Foreboding? Hesitation? That Josephine felt unsteady in front of a camera—a feeling that many of us have felt ourselves—should that lead me to conclude that her life in general was filled with trepidation and anxiety? Clearly not. Lives are not like that—they are a mix and Josephine undoubtedly had good days and bad. But then I come back to the message on the back of the card—"We are having a good time at the zoo"—and I cannot help but wonder. In the end, as animals of the hunt, of power and fear and worship, of work and value and entertainment, elephants walk through fraught territories in the history of ideas. It should not be much of a surprise that long after their deaths, their bodies—their bones and flesh, their hide, feet, tails, and tusks, and even the stories about them—rest in these territories as well.

CHAPTER TWO

Afraid of Mice

Josephine's tusks are kept in a large locked cabinet with three shelves that slide out like drawers. The shelves hold elephant parts, some retained for scientific or historical reasons and some probably simply because throwing them out seems somehow wrong. The top shelf holds a half dozen hollowed-out feet along with a shriveled trunk; the bottom shelf is stacked with parts of skins; and the middle shelf contains an assortment of molars and smaller tusks, including Josephine's. The cabinets in this room do not typically hold cleaned bones; they are repositories for slowly rotting skins and other parts treated, often over a century ago, with deadly concoctions intended to slow decay and offer some protection from insects. When you open these modern industrial cabinets, the smell can be overwhelming; the odors of chemicals—old formulas along with newer repellents and insecticides—mingle nauseatingly with putrefaction.

The cabinet is one of a half dozen for elephant parts. Most of the shelves hold the skins of elephants. Sometimes there are just small pieces of skin—a square of the giant Fénykövi elephant, for example, the taxidermized mount of which stands as the awe-inspiring centerpiece of the Smithsonian's National Museum of Natural History. In other places, the shelves hold what appear to be entire skins, including those of elephants shot by Theodore Roosevelt during his 1909–10 postpresidential hunting safari in East Africa. Most of the elephant collection at the National Museum is of African elephants, and when one looks at the skins as a group, it is easy to see why many in the nineteenth century talked of races of elephants. The colors of the skins vary strikingly from rusty reds to pales and dark grays, and some are more hairy than others. Each is remarkably distinctive. There are parts other than skins in these cabinets, too. There is a mounted tail, straight and rigid as a staff, soles of feet, single ears, flattened skins of trunks, a decayed and shriveled foot with a tag that reads "Transferred from Div. Ethnology in 1954. Not to be condemned, exchanged, or given away." With so many bones and other parts of animals in this collection, I found it difficult at times to remember that each object came from a life, was part of the existence and experience—the history—of a particular individual. But then that connection would hit me again. In one cabinet with seven shelves of skins, for example, there is one from a male elephant shot on March 5, 1928, twenty miles north of Bor in Sudan by W. L. Brown. What is surprising about this skin is the way it has been folded and placed in the drawer: on top is the animal's face with an eye closed as if it were asleep (fig. 2.1). This is not just a piece of skin.

A mounted taxidermized skin becomes a character or idea composed to memorialize an event or tell a particular story: the final moments before the kill, the loving parent, the trophy, the unobserved animal at a waterhole, the taxonomical instance in a row of similar and different creatures positioned the same way for ease of comparison.[1] It is "the gorilla that attacked me," "the gorilla I shot," or "the gorilla with its family." In taxidermy, we are shown the skin of an animal placed over a form and into a story, and it is almost

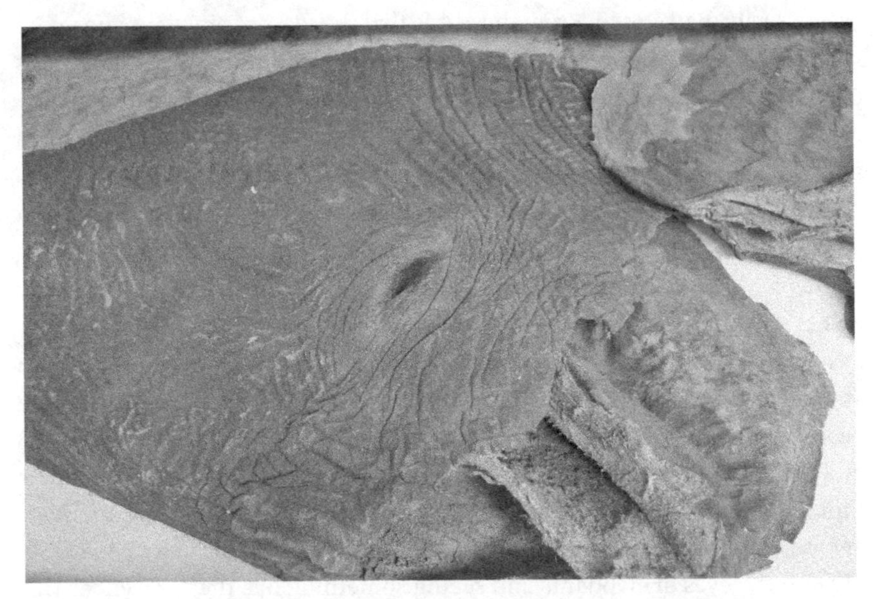

Figure 2.1. A face, National Museum of Natural History. Photograph by Nigel Rothfels.

impossible to conceive of a naked, unposed face. In zoos, too, and even in the wild, the pervasiveness and persuasiveness of narratives or stories, of ways to understand the animals—as endangered, as living free, as sleepy, as cute or terrifying—can make it difficult to grasp the vulnerable, naked life of an animal's face, to be confronted by its unembellished eloquence.

This unmounted skin in a drawer, a skin partially narrated in a tag telling who collected it, where, and when, was much more than an abstracted skin—it was a bare face.[2] The actual eye was missing, but in that closed lid there was undeniably, for me at least, the remembrance of an eye and a life. And what froze me was the skin around the elephant's closed eye. Very old beliefs about eyes and also fairly recent ideas about elephants and perhaps even nature help account for why I was drawn to the eye of the elephant in the way I was. First, the notion that eyes are the windows of the soul is an ancient one. In his *Natural History* written almost two thousand years ago, Pliny the Elder describes eyes in a way that seems very familiar to us today. "In all animals, there is no part in the whole body that is a stronger exponent of the feelings, . . . for it is from the expression of the eye that we detect clemency, moderation, compassion, hatred, love, sadness, and joy," he observes, adding that "beyond a doubt it is in the eyes that the mind has its abode: sometimes the look is ardent, sometimes fixed and steady, at other times the eyes are humid, and at others, again, half closed. From these it is that the tears of pity flow, and when we kiss them we seem to be touching the very soul."[3] Still, all the photographs of elephant and whale eyes in the popular media seem very modern, and my reaction in the museum, too, seems modern—it is as if these animals' eyes must somehow open to a deeper world of thoughts, of souls and dreams. Years ago, I read a letter on a memorial website about an elephant who died at a place in Tennessee called the Elephant Sanctuary. The letter from a volunteer relates that one day while he was painting fence posts, an elephant walked by on her way to the barn: "I will always remember looking into her eyes as she passed and the awe I felt . . . as if I was in the presence of some divine wisdom and grace, . . . which I believe I was. I don't have the words to describe the peace I felt being in the presence of these wonderful creatures. It remains one of my fondest memories of my 40 years thus far."[4] I have heard similar thoughts from so many people over the years I have been thinking about elephants, and I am left with a historical question: how long have we been looking into the eyes of elephants and seeing something like the "presence" this volunteer described?

So Great a Stranger

Ideas about elephants are a weave of old and new. Disney's 1941 *Dumbo*, for example, deploys a story that can be traced back at least to the Roman Empire—that elephants are afraid of mice—alongside more recent stereotypes of women in its depiction of female elephants screaming in fear at the sight of Timothy Mouse. If you spend any time in elephant barns, you will realize quickly enough that the buildings can provide habitat for quite a number of birds and small mammals, as well as invertebrates, who scurry about or spin webs in corners, and you will also see that generally, elephants do not seem to care about them one way or the other. Trying to discredit ancient stories of elephants, though, is like playing that old game whack-a-mole—as soon as you think you have put one legend to rest, another pops up, and before long the first one shows its head again. The fact that ancient stories about elephants such as that they are afraid of mice or mourn their dead have persisted does not mean, though, that when we see elephants today we are experiencing them and thinking about them in the same ways as in the past. Our ideas about elephants have changed over the centuries. Even a quick look, for example, at medieval European and early modern notions about elephants makes clear that a European five hundred years ago could not have imagined the kind of encounter with an elephant that the volunteer wrote about.

In our thinking about these elephants from longer ago, we first need to recognize that although individual elephants did show up fairly regularly in medieval and early modern Europe—as many as two or three every century—the animals never lived long and were seen by very few. Even though transporting the animals typically meant walking with them from town to town until they reached their final destination at a royal or other private collection, only an exceedingly small number of people could expect even the briefest glimpse at an elephant in their lives. In the preamble to a small English pamphlet from 1675 entitled "A True and Perfect Description of the Strange and Wonderful Elephant Sent from the East-Indies," the anonymous author describes the elephant as "so great a stranger in these parts; there having never been but one of them before in England; so that very few Persons now alive amongst us, but such as have Travelled the Eastern World, ever saw one of them, unless upon a Sign-Post in wretched painting."[5] The author is correct about the rarity of elephants to that point in England; more than four hundred years separated the arrival of this elephant from the one that had preceded it as a gift to Henry III in 1255. The author is also correct, though, that many people knew of the existence of elephants, even if their understanding of the animal came through "wretched" illustrations.

Elephants were not like cows, horses, pigs, chickens, crickets, goats, spiders, pigeons, trout, rats, mice, or even boars, deer, or wolves; for almost everyone, elephants were among the fantastic beasts—lions, tigers, dragons, unicorns, rhinoceroses, satyrs, hyenas, apes, and many more—that one would never see but would occasionally hear about or see in illustrations. Elephants were described in scholarly works.[6] These accounts were passed down, copied, abridged, and expanded on for centuries, but they could only be read by the small number of people who could access the rare manuscript copies. They were the sources, though, for the often richly illustrated and more broadly available manuscripts collectively known as "bestiaries" that began to be produced in the tenth century. The bestiaries combined natural histories with moralizations—explanations for how the lives of animals revealed the presence of God, echoed the stories of the Bible, or simply reflected the banal lessons of daily life. But even the bestiaries were not that common. As the art historian Willene B. Clark notes, there are fewer than 150 known surviving bestiary manuscripts in Latin and the vernacular languages of Europe.[7] Nevertheless, it is clear that much of what is presented in these works was disseminated across Europe through oral means that included performances, sermons, and stories. The bestiary account, for example, of how the song of the nightingale (*luscinia*) helps us through the dark nights, and how that song also recalls a mother easing the sorrows and poverty of her children through her own songs, is the sort of story that gets shared by a priest or by a wise person in a village and then gets passed forward for generations.

In the end these texts were reflective of broader thought in the period despite the fact that most people did not have direct access to manuscript copies of them. The books presented what was known about elephants, pelicans, phoenixes, and mandrakes, and they also provided instruction in how to understand the world, God's creation, and much more. While people may have been utterly dumbfounded at the descriptions of elephants traveling through the countryside with their masters, they may also have recalled parts of accounts of the creature that they had once heard or remembered images they had seen painted on church ceilings, carved into pews, or illustrated on signposts. Those accounts and images were often very far removed from what we would recognize as the results of direct observation, but in the absence of face-to-face interactions with the animals, fantastic depictions could persist for centuries. When, at the beginning of the seventeenth century, John

Donne observes of the elephant that "nature hath given him no knees to bend" and Shakespeare has Ulysses exclaim, "The elephant hath joints, but none for courtesy," they are both referring to a story about elephants lacking knee joints that became widely known in Europe in the Middle Ages. James Thomson's *The Seasons*, an early eighteenth-century poem about nature, alludes again to the same story and that, consequently, elephants sleep while leaning on trees. In *Summer* (1730) he writes,

> Peaceful beneath primeval trees that cast
> Their ample shade o'er Niger's yellow stream,
> And where the Ganges rolls his sacred wave,
> Or mid the central depth of blackening woods
> High-rais'd in solemn theatre around,
> Leans the huge elephant—wisest of brutes!
> Oh truly wise! with gentle might endow'd,
> Though powerful, not destructive.[8]

While ancient authors like Aristotle and Aelian had noted the difficulties elephants have in lying down and getting up, the idea that elephants' legs were incapable of bending only began to circulate in late antiquity, making one of its earliest appearances in Cassiodorus's letters in the sixth century, over a thousand years before Shakespeare and Donne. The story managed to survive and spread because it made sense for the period; it is just one of many, many stories about elephants that were retained because they were very difficult to disprove and because they could help explain important aspects of the world.

In what became its more or less standard form, the account of elephants in the bestiaries was about a thousand words long, making it one of the longer entries in the work. It presented a compilation of ideas accumulated and winnowed over fifteen hundred years, stretching back as far as Aristotle and including descriptions from later authors such as Pliny, Aelian, St. Ambrose, and Isidore of Seville. The account provides a brief physical description of the animal, discusses its use in war carrying a howdah or even a fortified tower with archers or lancers on its back (fig. 2.2), notes that it is afraid of mice, and lives for three hundred years. It explains that the elephant is naturally chaste, that it can only be induced to mate after it consumes a rare fruit in secret in a special forest in the East, that it mates only once in its life and gives birth in water to protect its young from serpents or dragons. It claims that the elephant lacks knee joints and that it therefore cannot get back up after it has fallen but adds that if an elephant falls, it calls out and others quickly come to its aid; first one large elephant, then twelve others, and then, finally, one small elephant, who, because he is so small, is able to get under the fallen elephant and lift him up. It argues that although elephants are powerful and can be deadly, they are naturally gentle, thoughtful, intelligent, and just. Using these observations, the account then connects a series of lessons from the Bible, including stories about Adam and Eve and the fruit of knowledge, Moses (the large elephant who comes to save the fallen one), the Old Testament prophets (the twelve elephants who come after the big one), and Christ (the small and humble elephant who is able, finally, to lift the fallen elephant). Through these connections to the Bible, the "natural history" of the elephant is shown to reflect Christian teachings.[9] Even if people were not familiar with the full account of the elephant in the bestiaries, bits and pieces undoubtedly circulated widely. The image of an elephant with a tower or castle on its back was a popular decoration in and on churches. Elephants were known to be large, though some images showed them to be about the size of a large pig, while most suggested they were about the size of a horse. They were also known to have tusks (depicted variously as pointing up from the lower jaw, pointing down from the upper jaw, and emerging from both jaws), legs like columns, ears of all sizes, and, most distinctively, a trunk.

Ideas about elephants did begin to change slowly from the fifteenth through seventeenth centuries, partly because more elephants began to appear in Europe and partly because more ancient Greek and Latin texts describing the animals became available. Many of the ideas presented in the bestiaries, though, endured for centuries more.[10] For example, while in the 1550s Conrad Gessner strove in his monumental *Historia animalium* to use either his own observations or those of his correspondents, his lengthy account of elephants exhibits more of his skills as a scholar of texts than expertise as a proto-zoologist. Gessner's work, too, quickly became authoritative and was copied, translated, excerpted, abridged, and otherwise modified for another couple of centuries, despite the increasing presence of elephants that could be used to verify the often-ancient claims.

One highly influential example of the continuing spread of much older ideas via Gessner's work is the 1607 *Historie of Foure-Footed Beasts*, by the Englishman and contemporary of Shakespeare and Donne, Edward Topsell, largely a translation of *Historia animalium* with additions by Topsell intended to emphasize the importance of God and Christian teachings. Gessner, Topsell, and others did challenge some tenacious ideas about elephants. "It is false," Topsell writes, that elephants "haue no ioynts or articles in their legs,

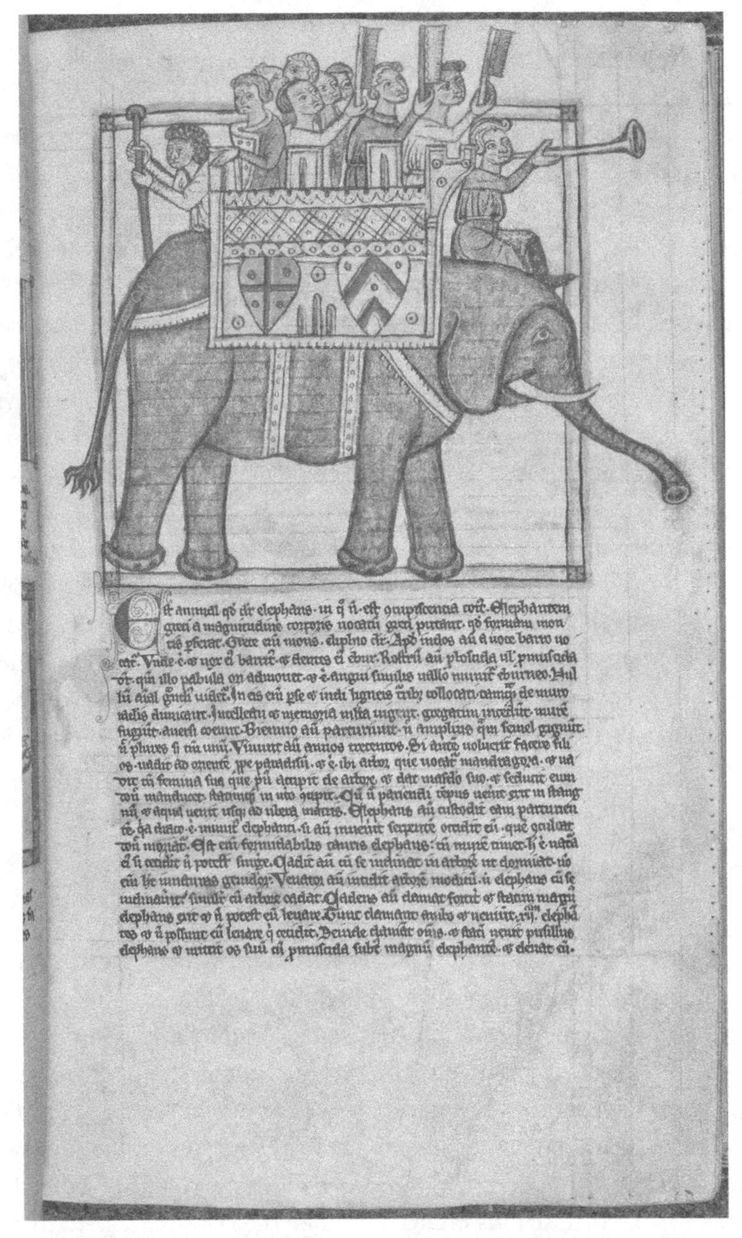

Figure 2.2. Entry for the elephant in a mid-thirteenth-century Latin Bestiary. © British Library Board, Harley MS 3244 f039r.

for when they please they can vse, bend, and moue them, but after they grow old, they vse not to lie downe or straine them by reason of their great weight, but take their rest leaning to a tree."[11] But if joints returned to legs, other ideas were reinforced. Most of the arguments, for example, about elephant mating behaviors are repeated by Gessner and Topsell.[12] Not surprisingly, certain ancient claims about elephants that had disappeared in medieval accounts reappeared in Gessner's account and in the accounts of those who followed him. Revising Pliny's claims, for example, Gessner and Topsell argue that elephants use the folds of their skin to trap and kill the flies that bother them and that they worship the sun and moon.[13] They also note that in proportion to their heads, the eyes and ears of elephants are diminutive. Topsell writes: "Their head is very great, and the head of a man may as easily enter into their mouth, as a finger into the mouth of a Dog; but yet their eares and eyes are not aequiualent to the residew of their proportion: for they are smal." Here, Topsell also includes a description that Gessner attributes to the first-century BCE author Varro: "Their eyes are like the eyes of Swine, but very red."[14] After over fifteen centuries during which the eyes of elephants were not discussed in Western texts about the animals, references to pig-like eyes begin to appear in the sixteenth and seventeenth centuries. It is unlikely, of course, that the eyes of elephants were never noticed or thought about during those many years. Rather, it seems that in comparison to tusks, trunks, jointless legs, overall size, and a series of surprising abilities and behaviors, elephant eyes were simply considered unremarkable.

The First and Grandest of Terrestrial Creatures

In the wake of ancient and medieval accounts of elephants, one of the most extraordinary texts in the whole history of natural history was published: Buffon's *Natural History, General and Particular* (*Histoire naturelle, générale et particulière*). Georges Louis Leclerc, Comte de Buffon, stands among a small group including Aristotle, Pliny, and Darwin as one of the most influential voices in Western thought on the subject of the natural world. From 1749, when the first three volumes of his natural histories of quadrupeds, birds, and minerals appeared, until long after 1789, when the last supplement to his encyclopedia was published, Buffon's *Natural History* remained the world authority on animals. When Thomas Jefferson referred in his *Notes on Virginia* to Buffon as "the most learned . . . of all others in the science of animal history," he was echoing a consensus opinion.[15] Available in a wide range of editions and abridgements, sizes, and prices, with and without its scientific illustrations, translated, copied, paraphrased, and quoted in publications for centuries, Buffon's work was read for knowledge, for instruction, for pleasure, and for moral edification, as well as because its sweeping plan suggested that he would end up writing about everything. After making their way through the whole of an entry in the *Natural History,* readers could be confident that they knew everything that needed to be known about that animal. Buffon was not, however, trying to write modern scientific descriptions from which he would omit all the anecdotes and allegories of the past centuries. For Buffon and his readers, part of the pleasure of the *Natural History* was to be found in the stories that brought the animal closer to human experience.[16]

Buffon begins his entry on elephants by expanding on Linnaeus's claim that the animals closest to man were the apes. For Buffon, four animals were deserving of special note. The ape obviously had a physical structure most like that of humans, but Buffon felt the dog should be lauded for its ability to form attachments based on affection. Beside these two, the beaver had to be recognized for its building projects that demonstrated an ability to cooperate in work and to form something like societies. Buffon insists, though, that the elephant, what he called the "first and grandest of terrestrial creatures," was superior to the other three because "in him all their most exalted qualities are united." The elephant, he concludes, has the advantages of the apes' hands, the dog's emotions, and the beaver's intelligence, combined with "the peculiar advantages of strength, largeness, and long duration of life."[17] According to Buffon, the elephant's extraordinary strength and size equipped the animal to battle lions, uproot trees, knock down walls, carry towers on its back, and lift burdens that a half dozen horses could not move. The very "earth shakes under his feet," Buffon writes, but the elephant also demonstrates "courage, prudence, coolness, and punctual obedience." Although elephants could be passionate in their mating and violent when attacked, the animals somehow seemed always moderate and judicious in their actions. The elephant, Buffon concludes, "is universally beloved, because all animals respect, and none have any reason to fear him."[18] Buffon describes the embodiment of intelligence, power, temperance, and discretion, all brought together in a gigantic vegetarian. He does not discard every controversial idea from the past, though. He accepts that elephants live for two hundred or more years, that the young suckle with their trunks, that elephants mate face-to-face, that the smell of hogs frightens them, and that they are naturally modest and refuse to breed while observed. But he may have accepted these ideas precisely because they fit so well within his overall conception of the elephant as the most remarkable of all animals.

While Buffon is describing the elephant, he also seems to be making arguments about how human society could be improved. Noting that elephants, for example, live in just and compassionate societies and are rarely seen alone, he insists that the animals are naturally gentle and use their strength and tusks only to defend themselves and their companions. When they are moving through the forests, the strongest stays in front, the next strongest brings up the rear, and "the young and the feeble are placed in the middle. The mothers carry their young firmly embraced in their trunks."[19] When it comes time to mate, the animals know "better than the human race, how to enjoy pleasure in secret" and seek out "the deepest solitudes of the woods, that they may give full vent, without disturbance or reserve, to all the impulses of Nature."[20] During a time, too, when both the domestication of animals and human slavery were topics of debate, it seems to have made complete sense to Buffon that elephants would so abhor captivity that they would suppress their most fervent desires in order to avoid perpetuating the slavery of their kind. Buffon writes, "They are not of the number of those born slaves, which we propagate, mutilate, or multiply, purely to answer our own purposes." While the elephant would accept its own slavery, it would not mate in captivity and thereby increase the wealth of its owner. "This circumstance," Buffon argues, "shows the elephant to be endowed with sentiments superior to the common brutes."[21] Despite the loss of its freedom, the elephant, Buffon claims, is both wise and moderate enough to accept it. He insists that the elephant is "the most gentle and most obedient of all domestic animals. He is so fond of his keeper, that he caresses him, and anticipates his commands, by foreseeing everything that will please him." He "never mistakes the voice of his master" and he fulfills commands with "prudence and eagerness." The elephant's character, Buffon indicates, "seems to partake of the gravity of his mass."[22]

Like most of the writers who preceded him, Buffon did not base his account of the elephant on personal observation. Rather, he consulted accounts passed down through history—those provided by Aristotle, Pliny, Gessner, and also Georg Christoph Petri von Hartenfels, whose less well-known but comprehensive summation of what was known about elephants, *Elephantographia curiosa*, was published in Germany in 1715. He also turned to reports he had received from travelers who had seen (or claimed to have seen) elephants in their range countries. More than anything, Buffon sought to distinguish the plausible and verified from the fanciful or unreasonable.[23] Using what he learned, he extended his arguments to logical conclusions. Because elephants had never been observed mating, for example, he accepts stories of modesty and builds them into a more comprehensive account of what had never been seen.[24] In the end, most of what Buffon had to say about elephants was quite old. Buffon did not mention mice, jointless legs, or giving birth in water to protect newborns from dragons, but most of his account, including that elephants were intelligent, forgot neither a good deed nor an injustice, were exceptionally powerful but also gentle, could as easily pick up a tree as a coin, were led by the most powerful, and lived for centuries, is familiar.

While the information in Buffon's entry for the elephant comes from quite old sources, however, the overall tone of the description, with its enthusiasm for reason, passion, prudence, and justice, is characteristic of the late seventeenth and eighteenth centuries. There are moments in which Buffon does relate a relatively new bit of information about the animals. He notes, for example, that elephants enjoy wine and distilled spirits, including the Indian drink arrack, and enjoy the smoke of tobacco—neither arrack nor tobacco were known in medieval Europe.[25] What makes Buffon's discussion a milestone, however, is his description of elephant eyes. Picking up on the idea that had appeared in the work of Gessner and other sixteenth-century writers that the eyes of elephants were disproportionate to their heads, Buffon describes the eyes of elephants in a way that seems strikingly familiar from our twenty-first-century perspective. He writes that though the elephant's eyes are small, they reveal a "pathetic expression of sentiment, and an almost rational management of all their actions." Comparing them to the eyes of dogs, which he also finds expressive, he insists that dogs' eyes move too quickly for us to distinguish the "successive shades" of the animals' thoughts but that because the elephant is "naturally grave and moderate, we read in his eyes, whose movements are slow, the order and succession of his internal affections."[26] By "pathetic expression of sentiment" ("l'expression pathétique du sentiment") Buffon means that elephant eyes communicate emotional states, that the eyes reveal the animals' feelings. They also reveal the overall calmness, seriousness, and even rationality of their minds. In particular, Buffon points to the friendliness and deliberateness of the elephant's gaze when it is attending to the instructions of its master.[27]

If soldier-filled towers and fights with dragons were among the most popular ways to represent elephants in the Middle Ages, the expressive eye described by Buffon marks the appearance of a very different entity. For Buffon and his readers, the elephant was no longer a terrifying magical beast. By the eighteenth century, the elephant had ceased to be something only of legends and was becoming the most admirable of animals, a creature known for its depth of feeling, its rationality, its strength and gentleness, its moderation, its commitment to family, and its sense for justice. Part of this change was no doubt due to the increasing presence of actual elephants in Europe, including over a half dozen in the seventeenth century and another dozen in the eighteenth, but, more contact with real elephants notwithstanding, the elephants described by Buffon, the elephants that people knew before they ever encountered any living specimen, were still much more creations of a history of ideas than anything else.[28]

The Corral

The great naturalists of the late eighteenth century and into the second half of the nineteenth were generally less interested in the lives of animals than Buffon. Rather, they became preoccupied with questions about the larger history of all life, how species were related to each other, and whether species could evolve into new forms. They were also trying to incorporate fossil evidence into their work that showed that species had apparently died out in the deep past. While the great French anatomist Georges Cuvier, for example, made the ground-breaking discovery in 1796 that mammoths were an extinct species, making them the first prehistoric animal, he added little in his influential 1817 *Le règne animal* (*The Animal Kingdom*) to what was known about the lives of elephants beyond insisting that they were no more intelligent than dogs.[29] Still, if most nineteenth-century scientists interested in animals focused on comparative anatomy, there were others who remained interested in how animals actually lived their lives, and they directed their work to increasingly large popular audiences.[30] Perhaps none of these scientists became better known than the German naturalist Alfred Edmund Brehm.

Born in 1829, the son of a village pastor with a reputation for his writings about European birds, Alfred Brehm became interested in animals in his late teens, joining a natural history expedition to North Africa before he went on to study natural sciences at the University of Jena. After completing his studies and further travel, Brehm found an income and increasing notoriety writing about animals in popular magazines destined for the German middle class. In 1860, at the age of thirty-one, he signed an agreement with a publisher to write the text for a richly illustrated encyclopedia of animals that would forego long discussions of comparative anatomy and taxonomy and instead focus on illuminating the lives of animals in the wild. The first edition in six volumes was published between 1864 and 1869 under the title *Illustrirtes Thierleben* (literally, "illustrated life of animals") and was a publishing success. Soon Brehm began preparing a second expanded edition, which appeared in ten volumes between 1876 and 1879 with the title it has carried through the many editions ever since, *Brehms Thierleben*, or, as it was translated into English, *Brehm's Life of Animals*.[31]

Brehm's entry for elephants in the first edition is thirty-four pages long and includes two full-page illustrations of African and Asian elephants by Robert Kretschmer, illustrations that, though not drawn in the field from life nevertheless tried to show the animals interacting with natural environments. In the illustration of the African elephant (fig. 2.3), for example, a small herd is shown browsing in a lush valley overlooked by sharp crags. Candelabra trees are prominent, though they are more likely to be found in dry savannahs than forests, and the animals themselves exhibit a mix of Asian and African elephant characteristics. The faces of the animals are not as broad or massive as one would expect for African elephants, and the bodies are depicted with the convex backs of Asian elephants. The ears reflect an awareness that African elephant ears can be very large, but the way they are affixed to and conform around the head and the way they meet over the neck is not realistic. However unrealistic the scene and however apparent it is that the artist had not had the opportunity to study living African savannah elephants, it is clear that the goal of the image is to show the animals as they might be seen in the wild, or at least how they were imagined in the wild.

In the text proper, Brehm, like those before him, relies on previously published works. After generally noting the geographic distribution of the species of living elephants and the kinds of environments they inhabit, Brehm turns to describing the lives of the animals. Because Brehm had access to reports in scientific journals, the growing literature being produced by hunters, and full-length monographs describing colonized countries, his sources were more carefully curated and substantial than what Buffon had at his disposal. Brehm doesn't retell the long history of ideas about elephants or dwell on towers, mice, dragons, innate feelings for justice, giving birth in water, or eating unusual fruits before mating. He does, however, address several questions still unanswered at the time, including whether elephants breed in captivity, require seclusion to mate, live as long as two to three hundred years, and how big they might grow.

To address these lingering issues, Brehm first turns to a paper published in the *Philosophical Transactions of the Royal Society of London* in 1799 by John

Figure 2.3. Robert Kretschmer, African elephant, in Alfred Edmund Brehm, *Illustrirtes Thierleben* (Hildburghausen: Bibliographisches Institut, 1865).

Corse entitled "Observations on the Manners, Habits, and Natural History, of the Elephant."[32] Corse spent ten years in Tripura, Bengal, where from 1792 to 1797 he was in charge of elephant-hunting operations for the government. During those years, Corse explains, he came to realize that much of what had been written about elephants in Europe was in error. In particular, he sought to dispense with the old stories of elephant modesty and Buffon's claim that elephants refuse to breed in captivity and so limit enslavement to themselves.[33] Reporting at length on his own observations of large numbers of people watching captive elephants mating, for example, he felt he had proven "the inclination of the elephant, to propagate his species in a domestic state, and that without any signs of modesty."[34] Corse was also convinced that both the height and intelligence of elephants had been exaggerated, and he presents data on scores of elephants he himself measured, demonstrating that most female elephants in his experience grew to between seven and eight feet at the shoulder and that most males grew to between eight and ten feet. Corse closes his paper with a quite modern point of the importance of adding to knowledge rather than just repeating what has already been reported. "In the course of this narrative," he states, "I have, in general, only related only such particulars concerning the elephant as came within my own knowledge, and which were, either not known, or not published."[35]

Corse was in many ways an ideal observer for Brehm. Rather than just a person traveling in India who happened to see elephants, he was a scientific thinker—someone who could recognize that there were errors in the existing literature and who then had the opportunity, inclination, and skills to correct those errors methodically. Moreover, his results were published in a prestigious journal that reached international audiences. But Corse's account was limited—it was neither particularly compelling to read nor wide ranging in its scope. To answer other questions about the lives of elephants, Brehm turned to a second and quite different kind of source: a former colonial secretary of Ceylon, James Emerson Tennent.[36] Tennent was born in 1804 in Belfast, and after his education and a stint in Greece, he served for five controversial years, from 1845 to 1850, in Ceylon (Sri Lanka). During his time on the island, Tennent explored widely and became a collector of natural history specimens and archaeological artifacts. He also studied the history of the cultures of the island using both European and local sources. His political ambitions in traveling to Ceylon appear to have been largely frustrated, however, because of conflicts with other colonial officials. After returning to the United Kingdom, though, he published a series of books, beginning in 1859 with his two-volume *Ceylon: An Account of the Island Physical, Historical, and Topographical* that quickly established him as the recognized authority on the island, its peoples, history, and natural history.[37]

Tennent was not a scientist but essentially a travel writer who prepared a thirteen-hundred-page account of a relatively small country he had lived in for five years. *Ceylon*, though, was an unusually successful book, going into six editions in its first year and quickly setting a standard that other colonial officials sought to emulate. The writing was thorough, perceptive, occasionally poetic, and often deeply sympathetic to the foreign landscape. Tennent was not, it seems, popular with settler colonists in Ceylon, but his descriptions of the outpost in the Indian Ocean appear to have captivated readers back home.[38] Part of what Brehm gleaned about elephants from Tennent's study was that at least one captive elephant in Ceylon had reached a documented age of 140 years. Brehm also includes Tennent's statistics of the rate of death over a twenty-year period of 138 elephants captured for the government—only one of the animals survived the full twenty years. He notes that Tennent observes that elephant herds were of varying size but were occasionally as large as a hundred individuals. Members of the herd, according to Tennent, were closely related and often shared certain physical traits like the color of their eyes. The herds were led, he claims, by the strongest animal, regardless of gender. Tennent also points out that when occasionally a member of the herd was either driven out or became separated from its group, it might then become violent and dangerous, turning into what is called a "rogue."[39] Elephants prefer the forest and shade, he maintains, but they can be found in diverse landscapes, and their paths invariably lead to the best passes across mountain ranges. Although wild elephants have little need of great intelligence in Ceylon, because they have unlimited quantities of food and no enemies save man, tame and trained elephants, he argues, can exhibit remarkable reasoning powers.

From what we know today it looks like much of what Tennent reported as having seen himself was more likely reported to him secondhand, but Brehm seems to have found the account essentially entirely credible.[40] What apparently impressed him most was Tennent's sense for the deeper character of elephants. Tennent was struck by the calm and an ostensibly utterly inoffensive demeanor of the animals in the wild. Calling the animals "essentially harmless" and describing them as only wanting to live peacefully beside all other animals, he concludes that "with the exception of man, *his greatest enemy is a fly!*"[41] In Brehm's and Tennent's view, the elephant is fundamentally a timid, retiring, thoughtful, and compassionate creature of the forest. For both, too, the elephant seems only to suffer at the hand of human beings: European hunters, on the one hand, were self-evidently killing the animals for no good reason, while others were engaging in what they both felt was one of the most compelling though also somehow deeply tragic practices: capturing wild elephants in corrals and taming them for work. Brehm finds Tennent's account of the corral (or *keddah*) so "enthralling and thorough," that he incorporates the entire forty-three-page chapter from *Ceylon* into his own text.[42]

Tennent begins this account by noting that before the arrival of Europeans to the island, only a small number of elephants were required for pageantry and processions and that these elephants were typically caught individually by specialized hunters. However, the colonial powers, the first of whom, the Portuguese, arrived in the sixteenth century, followed by the Dutch and the English, began staging large-scale captures because they realized that elephants could be critically important in clearing forests and constructing roads and other needed colonial infrastructure.[43] According to Tennent, Ceylon adapted an already-existing Indian method for capturing herds of elephants using a large concealed corral in the forest built with strong vertical timbers connected by crossbeams and braced by other beams on the outside of the structure. He describes at length a corral he witnessed near Kornegalle in 1847. Setting the scene, Tennent emphasizes the beauty and peacefulness of the ancient forest: the path to the scene "led across open glades of park-like verdure and beauty, and at last entered the great forest under the shade of ancient trees wreathed to their crowns with climbing plants and festooned by natural garlands of convolvulus and orchids." "Here," he explains, "silence reigned, disturbed only by the murmuring hum of glittering insects, or the shrill clamour of the plum-headed parroquet and the flute-like calls of the golden oriole."[44]

According to Tennent, the site was carefully chosen because of the presence of nearby herds and because of natural features in the landscape, including a river that the animals could use for watering after they had been captured. The structure hidden in the forest was about 500 feet long and 250 feet wide and was built by "vast numbers" of locals over the course of many weeks.[45] One end was left open with bars that could be slid into place to close it up. From there, two long arms of fencing reached out into the forest beyond the corral forming a funnel. When all was set, two to three thousand beaters moved out into the forest to enclose an area miles across. Once they were in position, they began slowly and quietly to move the animals toward the corral over the course of several days. According to Tennent, the drivers took advantage of the elephants' natural "timidity and love of seclusion"—only a slight disturbance, he argued, would encourage the animals to slowly move in the desired direction toward the corral.[46] At last, after two months of preparations, it was time and, when a signal was given, the final drive of the elephants was made in the night; the beaters pushed the elephants into the corral by burning torches and making loud noises. Tennent describes the final tense moments when the elephants "approached at a rapid pace, trampling down the brushwood and crushing the dry branches." The leader of the herd paused briefly at the gate and then rushed in followed by the herd. At that moment, the forest lit up "with a thousand lights" as every hunter rushed "forward to the stockade with a torch kindled at the nearest watch-fire."[47]

The animals had been captured. More animals were brought in the next day, and then came the task of securing each of the captives. At this stage, two trained elephants entered the corral carrying their mahouts and men with ropes. According to Tennent, one of the elephants being used by the captors had been "in the service of the Dutch and English Governments in succession for upwards of a century"; the other, called Siribeddi, was about fifty years old and distinguished for her gentleness and docility."[48] Siribeddi was particularly adept in her role, Tennent states. She moved quietly and slowly, "with a sly composure and an assumed air of easy indifference" toward the herd of captives, halting every now and then to eat some grasses. When the herd noticed her, the leader approached and "passed his trunk gently over her head." The leader then returned to his herd, and Siribeddi followed while one of the men with ropes slid off her back and put a noose around one of the leaders' rear legs.[49] The other end of the rope was pulled around a strong tree and the two tame elephants pulled the ropes until the wild elephant had been secured close to the tree. Tennent reports that the tame elephants seemed to exhibit "the utmost enjoyment in what was going on," while their "caution was as remarkable as their sagacity."[50]

The circumstance of the captives, on the other hand, struck him as filled with pathos. In the text that accompanies an illustration by the artist Joseph Wolf of a tusked male heaving against its ropes while grasping a limb with its trunk (fig. 2.4), Tennent explains that each of the elephants responded to its circumstances differently. Some offered little resistance, while others "dashed themselves on the ground with a force sufficient to destroy any weaker animal." Some were quiet in their struggles, and others "bellowed and trumpeted furiously, then uttered short convulsive screams, and at last, exhausted and hopeless, gave vent to their anguish in low and piteous moanings." Some elephants just lay motionless on the ground with no other sign of their "suffering than the tears which suffused their eyes and flowed incessantly."[51] In describing a lone tusker, not part of a herd, that had been captured in the operation, Tennent notes that "when overpowered and made fast, his grief was most affecting; his violence sunk to utter prostration, and he lay on the

Figure 2.4. Joseph Wolf, elephant corral, in James Emerson Tennent, *Ceylon: An Account of the Island Physical* (London: Longman, Green, Longman, and Roberts, 1859). American Geographical Society Library, University of Wisconsin–Milwaukee Libraries.

ground, uttering choking cries, with tears trickling down his cheeks."[52] These elephants, straining against ropes and in anguish on the ground, victims of a treachery, are something new.

By the time Tennent and Brehm wrote their accounts, Buffon's elephants, the elephants of the medieval bestiaries and Pliny the Elder, the elephants who were afraid of mice, were long gone. Tennent's elephants, with tears creasing from their eyes, are wild, dynamic, and vividly described, and yet it is also clear they are being seen through the at times conflicted lenses of nineteenth-century colonialism. Tennent captures the century's fascination with the colonies—places where life seemed both more dangerous and more spectacular—while also suggesting the growing awareness that the presence of Europeans in the colonies signaled the end of an innocence. In Brehm's and Tennent's descriptions, there is often a hint or more of remorse, a sense that with all the progress, with the expansions of colonies and knowledge, there is an inevitable loss. The account of the corral with its faraway veneer intruding on the pristine and idyllic forest points simultaneously to the accomplishments of European expansion and the loss of the primeval and unspoiled. Siribeddi, with her enthusiasm for participating in the capture of other elephants, can be seen as both an elephant and a stand-in for the landscape and cultures of Ceylon, as she literally betrays her kind and the island itself in helping the British build their roads and expand their plantations.

Unendingly Melancholy

When Brehm lightly revised his entry on elephants for the second edition of his encyclopedia, he closed his account by relating a story about elephants in Africa from the recently published memoir of German botanist and explorer Georg Schweinfurth. From 1868 to 1871 Schweinfurth traveled through central Africa, entering from the north rather than from the east or west coasts, which was more typical. In early 1869 he set out from Khartoum and headed south along the White Nile before turning west, moving into what is now South Sudan and following the Bahr el Ghazal river until he ended up in what is now the Democratic Republic of Congo. During much of this journey, Schweinfurth attached himself to a group of ivory hunters and dealers, although his memoir focuses more on the peoples, plants, and geography of the region than on the ivory trade itself. Despite his friendship with his traveling companions, Schweinfurth's perspective on the trade is highly critical. Calling the trade a "war of extermination" in his 1874 two-volume *Im Herzen von Afrika* (translated as *In the Heart of Africa*), he points to what motivates the trade: "Our walking-stick handles, billiard balls, piano keys, our combs and fans, and hundreds of such worthless things." Soon, he believed, the elephant would "join the ranks of those that have become extinct, such as the aurochs, Steller's sea cow, and the dodo."[53]

Near the end of his travels, the ivory traders presented Schweinfurth with a young elephant, whose mother they had recently killed, as a gift. Hoping to keep the young animal alive, Schweinfurth began feeding it milk from the cow he had brought with him so he could have access to fresh milk every day. After several days, though, the young animal died; Schweinfurth concluded the causes were insufficient food and the stresses of its experiences during the capture and march to Schweinfurth's camp. "For me," Schweinfurth recalls, "there was something unendingly melancholy in watching the already quite large yet so helpless creature die while breathing with such difficulty." "Whoever has looked into the eye of an elephant," he writes, "will find that, despite its diminutiveness and the short-sightedness with which these animals are born, that eye holds a more soulful look than that of any other quadruped."[54] Buffon's "expression of sentiment" in the eye of the elephant marked the beginning of a modern way of thinking about elephants. Schweinfurth's story, related in a German animal encyclopedia a century after Buffon, makes clear that by the middle of the nineteenth century, the elephant observed by the volunteer at the elephant sanctuary in Tennessee had fully entered the Western understanding of this animal.

In the last lines of his entry in the second edition, Brehm writes, "Still many herds of stately animals pass through the forests of Africa; but more and more they suffer from pursuing man. In the north and south, as much as in the coastal lands of the east and west, and even in the interior of Africa, the future is clear: they are to be crossed off from the lists of the living. In the upper Nile countries, where the ivory trade has been active for decades, they have already been completely exterminated, 'and it would not be difficult,' says Schweinfurth, 'to plot the entire area of the Bahr el Ghazal river over successive periods of five years to show how these animals have partly retreated before mass persecution and partly disappeared entirely.'"[55] For Brehm, killing elephants for their tusks was a fundamentally appalling act. He was not opposed to hunting for science or even for sport; indeed, he was a hunter his whole life. But the hunting of elephants bothered him. He seems to have been convinced that the only people who would shoot this remarkable animal were those interested simply in adding numbers to their hunting logs or ivory to their wagons.

CHAPTER THREE

A Serpent for a Hand

About thirty years ago I spent a few weeks looking through volumes of data collected by the Trade Statistics Bureau for the city of Hamburg, Germany.[1] Sitting in a library there, I went through sixty-four volumes covering the period from 1850 to 1913, looking for the amounts and values of a small list of imported raw goods. In the second half of the nineteenth century, Germany developed from an assortment of loosely confederated, inward-looking cities, states, ecclesiastical entities, and towns into an empire that included colonies around the world. I wanted to see if the importation of certain products in those years echoed the larger political, commercial, and cultural changes reshaping Germany. Among the items I tracked were seal pelts, tortoise shell, ostrich and other decorative feathers, whale oil, baleen, palm oil, hippopotamus and walrus teeth, and elephant ivory.

Overall, both the quantities of the materials imported into Hamburg every year and their value increased over the sixty-four-year period, reflecting the rapid growth in trade in the city more generally.[2] For example, in 1850 only 2,600 kilograms of turtle and tortoise shell were imported into the city, but by 1913, that amount had grown to 14,830 kilograms. In the same years, the quantity of whale oil imported rose from 3,739,300 to 51,767,600 kilograms, palm oil from 1,951,750 to 19,889,400 kilograms, and feathers from 1,050 to 89,152 kilograms (an almost eighty-five-fold increase).[3] In the case of elephant ivory, the quantity imported increased over the period from 61,650 kilograms in 1850 to 206,200 kilograms in 1913. Numbers like this appear abstract, so it is worth pausing for a moment to try to develop a concrete sense of what they mean. A pound of feathers weighs the same as a pound of anything else, but what would the almost 200,000 pounds (89,152 kilograms) of decorative feathers (not feathers used for bedding) brought into Hamburg in 1913 actually look like? And how many hundreds of millions of birds does it take to make 6,652,777 pounds of decorative feathers—the total imported over the sixty-four-year period I studied? Just what sort of network of killing, plucking, and shipping is required to account for all these birds and what kind of similar operation could account for the more than two million seal pelts brought into the port during the period?[4] And Hamburg was just one port city in Europe; as important as it was to Germany, it was a smaller port than New York or London.

From 1850 to 1913, over twenty-four million pounds of elephant ivory were imported into Hamburg, mostly through Great Britain and Portugal, although shipments directly from Africa increased in the last quarter of the century.[5] The average weight of tusks year to year varied a great deal. Hamburg kept statistics on the number of tusks (in addition to weights and values) for the years between 1850 and 1872, and during those years the average weight per tusk reached a high in 1865 of 41.54 kilograms and a low of 13.20 kilograms in 1869. Using a conservative average of 15 kilograms per tusk for the period from 1873 to 1913 and following the standard estimate of 1.88 tusks per elephant (not every elephant had two tusks), we can estimate that Hamburg imported the ivory of about 390,000 elephants between 1850 and 1913.[6]

In the fifteen years before the outbreak of World War I, Hamburg imported an average of 233,326 kilograms of ivory every year. Ian Parker, who conducted research on the subject in the 1970s, argues that at its height at the end of the nineteenth century, as much as eight hundred metric tonnes of ivory was exported out of Africa every year (so, a little less than four times the total imported by Hamburg alone)—the ivory of somewhere between thirty and fifty thousand elephants every year. These are staggering numbers and significantly higher rates than we have seen in recent years—in fact as much as two to three times our current rates.[7] But who was doing the killing and why? Answering the "who" question is perhaps the simpler task. From what we have learned since then, it seems the efforts of the so-called legendary elephant hunters of the nineteenth and early twentieth centuries only accounted for a small amount of the ivory imported into Europe or the United States.[8] While Walter Dalrymple Maitland "Karamojo" Bell claimed to have killed over a thousand elephants in Africa over his career and Thomas William Rogers over fourteen hundred in Sri Lanka, very few of the well-known hunters ever killed more than a couple of hundred of the animals. But if that is the case, how do we account for the millions of tons of ivory that reached European and American markets in the half century before the outbreak of World War I?

The simple answer is that ivory buyers, including slave traders and big-game hunters, encouraged indigenous people to collect ivory by providing them trade goods, including weapons.[9] In his memoir of his travels in Central Africa, Georg Schweinfurth describes some of the methods used to kill elephants. In one example, he reports that thousands of Azande worked together to drive herds of elephants into areas of the savannah with especially high and thick grass and then lit the grass on fire. Overcome by fire and smoke, the animals either died in the flames or were easily speared to death by the hunters. "In such a war of extermination," Schweinfurth writes, "not only are the males, who are armed with the large and valuable tusks, but also the females and the young are shamefully killed."[10] In the end, what drove the ivory market in the nineteenth century—and what continues to drive it today—is a combination of market demand outside the countries of origin and limited economic, health, legal, and educational supports and protections in the places where elephants can be found. In short, almost all the elephants killed for ivory have been killed by people with few options who are paid very small sums to provide tusks to a chain of procurers, traders, and merchants, each of whom earns more profit than the preceding link in the chain until the end product reaches a consumer. The reason for the ivory trade over the last thousands of years has always been the same: elephants are killed in Africa and India because their ivory tusks are in high demand in the Northern Hemisphere. People living in Western countries today may want to point fingers at Asian countries for continuing the trade, but every time an ivory heirloom is paraded on the *Antiques Roadshow* and the owner is told it is of great value, the underlying and very old forces of the trade are reinforced.

There are, nevertheless, differences between the ivory trade before the middle of the twentieth century and today. Most importantly, before the complete ascension of the use of plastics, ivory in the industrializing countries of the late nineteenth and early twentieth centuries was less an extreme luxury item and more of an industrial raw material used to produce a wide range of ordinary objects like piano keys, combs, handles of various kinds, billiard balls, and the handles of cutlery.[11] As a raw material, ivory for most consumers was almost entirely disconnected from the realities on which the industry was built. People who shop for meat in a supermarket today will know that while selecting a choice cut for dinner, they are unlikely to be thinking about industrialized slaughter in the food industry, and I think very few of those who held ivory-handled cutlery in the nineteenth century ever considered the elephant deaths hidden in those forks and knives.

Given the world-wide market demand, it is even a bit surprising that elephants have survived at all over the last few centuries. Many other species, of course, disappeared. Some were wiped out for meat or fat like the dodo, Steller's sea cow, several antelope species, and the passenger pigeon. Others were eliminated because they posed a perceived threat to people or livestock, including the wolf in England, Scotland, and Wales, quite a few species and subspecies of cats around the world, and the Tasmanian wolf or thylacine. A few species, like the ivory-billed woodpecker and the great auk, were even exterminated at least partly through the efforts of museums to collect the last ones. Despite intense hunting pressure, however, elephants managed to survive, and it seems that they did so at least partly because of ideas that developed and stories that were recounted about them in the nineteenth century. And some of the most important of these stories were told by hunters themselves. The reason international hunters today think it is entirely worth it to pay $30,000–$50,000 to legally kill an elephant in the wild is grounded in ideas about elephant hunting that emerged in the nineteenth and early twentieth centuries. Likewise, the origins of efforts to save elephants from extinction were these same stories by famous and sometimes infamous hunters that brought the realities of elephant death to the attention of a broader public. While efforts to save the thylacine came too late and fell on largely unsympathetic ears, stories of elephant hunting helped galvanize critics who believed that elephants should be protected from overexploitation, and some of those critics were, in the end, elephant hunters themselves.[12]

The Monarch

When Alfred Brehm turned to issue of elephant hunting in the first edition of his animal encyclopedia, the *Illustrirtes Thierleben* (1864–1869), he made clear that he found the practice a disgrace. He writes, "when I call the hunters infamous instead of famous, I have a good reason. Most of them are wholly unworthy of the hunt which they pursue."[13] Brehm reserves his special condemnation for the Scottish adventurer, writer, and lecturer Roualeyn Gordon-Cumming, whose two-volume account of his experiences, *Five Years of a Hunter's Life in the Far Interior of South Africa*, was published in 1850. One particular story Gordon-Cumming relates caught Brehm's attention, and he quotes almost the entire and quite long account.

The hunter was on horseback with his pack of dogs when about a hundred yards distant he saw "the tallest and largest bull elephant" he had ever seen. The "old fellow," Gordon-Cumming states, was distracted by the dogs when he fired. "Before the echo of the bullet" could reach his ear, he could see he had hit the elephant in the shoulder, "rendering him instantly dead lame." The dogs then descended on the wounded elephant, who, "finding himself incapacitated, . . . seemed determined to take it easy." "Limping slowly to a neighboring tree," Gordon-Cumming continues, the elephant "remained stationary, eyeing his pursuers with a resigned and philosophic air."[14] Realizing that the "noble elephant" could not escape, the hunter decided to "devote a short time to the contemplation" of him, so he gathered wood, started a fire, and prepared a kettle of coffee. "There I sat in my forest home," he recounts, "coolly sipping my coffee, with one of the finest elephants in Africa awaiting my pleasure beside a neighboring tree. It was, indeed, a striking scene; and as I gazed upon the stupendous veteran of the forest, I thought of the red deer which I loved to follow in my native land, and felt that, though the Fates had driven me to follow a more daring and arduous avocation in a distant land, it was a good exchange which I had made, for I was now a chief over boundless forests, which yielded unspeakably more noble and exciting sport."[15]

This is a "striking scene," but like all hunting stories, it is, before anything else, just a story. We'll never know if any of what Gordon-Cumming claims to have happened actually did, but we can know that it was important to him to paint this scene the way he has, filled with bucolic reverie and a reference to Scottish red deer—a reference that would undoubtedly have evoked Sir Edwin Landseer's *The Monarch of the Glen*, a painting completed in 1851, in the minds of a number of his readers at the time. Gordon-Cumming's account asks us to imagine that he was not hot and sweaty from riding, not agitated with adrenaline after having lamed the largest elephant he had ever seen, not aware of the large pack of dogs running around barking, not concerned with the fussing of his horse, not even seeing the presence of his guides and gun bearers but rather that he was sitting calmly sipping coffee in a solitude he wants us to think of as recalling the Scottish highlands and long-ago hunting adventures with deer.[16]

From our perspective in the twenty-first century, this sort of writing often seems preposterous, but most of Gordon-Cumming's readers at the time, including Brehm, did not see anything ridiculous about it. One reason for this is that the hunter was writing in the expected style for hunter-adventurers, as exemplified by William Cornwallis Harris, his predecessor in South Africa, whose 1839 *Wild Sports of Southern Africa* became an immediate classic. One can easily see the tradition which Gordon-Cumming follows in Harris's description of one of his first views of a large herd of elephants:

> Here a grand and magnificent panorama was before us, which beggars all description. The whole face of the landscape was actually covered with wild elephants. There could not have been fewer than three hundred within the scope of our vision. Every height and green knoll was dotted over with groups of them, whilst the bottom of the glen exhibited a dense and sable living mass—their colossal forms being at one moment partially concealed by the trees which they were disfiguring with giant strength; and at others seen majestically emerging into the open glades, bearing in their trunks the branches of trees with which they indolently protected themselves from flies. The back-ground was filled by a limited peep of the blue mountainous range, which here assumed a remarkably precipitous character, and completed a picture at once soul-stirring and sublime![17]

Harris made a watercolor of the scene (fig. 3.1), which he included along with twenty-five others as color plates in the 1852 edition of his book. In the foreground, a wounded elephant, bleeding from a dozen or so shots runs . . . it is not clear where . . . with its trunk held high in what appears more of a salute than a cry of pain. Further back, two hunters, out of scale and riding what appear to be carousel horses, are shown looking out over a valley filled with elephants who apparently have very little to eat, and, in the distance, the blue mountains loom. The whole scene, which is similar to contemporary depictions of bison herds in the American West, is imbued with the same light airiness that pervades Harris's prose. Harris is the sort of writer who describes situating himself in a wooded defile to await his quarry—in his writing, Africa is a place that is very far from that of the terrifying and gloomy dark continent that became the norm only decades later and is described so well in Joseph Conrad's *Heart of Darkness*, but this is precisely the kind of writing that Gordon-Cumming admired and emulated with his unhurried descriptions of hunts.

After having "admired the elephant for a considerable time" while enjoying his coffee, Gordon-Cumming decided to test the effect of various shots at the elephant's head. He explains that while he "fired several bullets at different parts of his enormous skull," they did not seem to affect the animal much. The elephant only "acknowledged the shots by a 'salaam-like' movement of his trunk, with the point of which he gently touched the wound with a striking and peculiar action."[18] Realizing—one has to say, *finally*—that he was "only tormenting and prolonging the sufferings of the noble beast," Gordon-Cumming decided it was time to finish off the wounded elephant as quickly

Figure 3.1. William Cornwallis Harris, "Hunting the Wild Elephant," in William Cornwallis Harris, *Wild Sports of Southern Africa* (rpt. 1852; London: John Murray, 1839).

as possible. He began a barrage on the elephant's left side, behind the shoulder, hoping to hit the animal's heart. After six shots with his "two-grooved" to which the elephant "evinced no visible distress," he fired three more shots with his larger "Dutch six-pounder." Finally, "large tears . . . trickled from his eyes, which he slowly shut and opened; his colossal frame quivered convulsively, and, falling on his side, he expired." Gordon-Cumming had the tusks cut out, which he reported were "beautifully arched" and "the heaviest" he had collected to that point, "averaging ninety pounds weight apiece."[19]

To this account, Alfred Brehm responds, "How infinitely higher stands the elephant over man, how wretched, how vile, the contemptible, treacherous enemy shows himself to be compared to the magnificent creature."[20] For Brehm, that Gordon-Cumming would, first, lame the elephant and then take his time considering the scene while the elephant suffered and that he would, second, experiment on the animal trying to gauge the success of various shots at his brain was simply unforgivable. Comparing Gordon-Cumming with royalty of the past who would have "hundreds of noble animals driven into a confined space and then would massacre them from an elevated stand" and other elephant hunters who shot "a good portion of their bags in the corrals and capture stations" "in order to enter a few more numbers into their disgraceful hunting registers," Brehm calls the unsportsmanlike behaviors of people like Gordon-Cumming "abominations."[21]

In calling out Gordon-Cumming, Brehm should not be seen, though, as expressing what had become a growing concern in the period about cruelty toward animals. Brehm simply believed that his audience—educated late nineteenth-century Germans—had (or should have) a different understanding of acceptable hunting behavior. The etiquette that Brehm expects of hunters comes into clearer focus in a small anonymously authored article entitled "Wounded Game" that appeared in the German bourgeois weekly *Die Gartenlaube* in 1895. According to the article, a "dark side" of hunting was that "every day thousands" of game animals "are wounded and finally die only after weeks of incredible suffering."[22] To make the point, the author relates a story that had recently been published in a German hunting magazine about a male roe deer that had been seen near a village. Upon hearing of the buck, a local hunter grabbed his gun and went to take a look. In the company of three others and two horses, he eventually encountered the animal walking slowly around in a field, about a hundred paces away. When they stopped to see whether it was possible to get a shot, the animal suddenly looked right at them and then slowly began walking directly toward them. The hunter recalled that he did not then shoot because he wanted to see what might happen. When the buck approached closer, though, he realized that "his lower jaw had been shot off and that it was hanging only by the skin." Still the hunter waited. "With almost every step" the buck "let out a broken moan and stared at us with complete misery." When the buck was thirty paces away, the hunter could bear it no longer and put the animal out of its misery. The hunter, at a loss for explaining the buck's behavior, speculates that "he approached our group of four men and two horses to find help."[23] In an admonitory tone, he concludes: "Even the hunter who has grown gray under the trees, who adheres strictly to the rules of using a gun, can never be certain that his shot will bring down his game. Hunting is absolutely an inexact craft! But he who would hunt as a huntsman, with all that this term entails, he has a heart for the game and accepts it as his bounden duty to shorten the suffering of the wounded by following its spoor."[24]

What made the story told by Gordon-Cumming so objectionable to Brehm was not that an elephant was killed but how it was killed. Brehm was appalled that Gordon-Cumming would take his time, build a fire, and sit and enjoy his coffee while the elephant, in pain, awaited the hunter's "pleasure," that he would experiment with shots and describe how the elephant would touch his wounds with the tip of his trunk, that even after deciding to put the animal out of its misery it took him another nine shots before the elephant finally closed his tear-filled eyes.[25] For people like Brehm, hunters had a moral obligation to kill the animal as quickly as possible.

These stories, and the responses to them, are anchored in the ever-evolving rules of hunting—the "rules of the game." People who grow up hunting or fishing—like the hunters discussed in this chapter—generally have strong views about what practices are right and wrong, even if their ideas might differ from the law or the ideas of their neighbors. Today, when it comes to hunting deer, for example, some people are comfortable driving or baiting and others find both of these practices objectionable; some people feel bow hunting unnecessarily increases the suffering of deer, while others highlight the heightened challenge of bow hunting and insist that using a bow is fairer to the animal than using high-powered rifles. The rules of hunting are inescapably historical and situational, personal and cultural, springing from traditions and lessons passed down within families and from one hunter to another. The rules for Gordon-Cumming and for Brehm were simply different. As rules rooted in culture, though, they are constantly being defined, defended, critiqued, and changed.

With that said, it is impossible for people who fish or hunt regularly to abide even by their own rules in every instance. In the end, there are so many unpredictable moments during hunts that mistakes are inevitably made. Hunters who write memoirs, of course, typically avoid describing occasions when they violate their own rules. At most, they will take a moment to explain what to others might appear like a violation by pointing to factors that made it impossible for them to do what they would normally have expected. *I would naturally have followed up on the wounded animal*, they might say, *were it not for the fact that I would certainly have died if I had tried.* Thus, when Gordon-Cumming describes his surprise to discover that his experiments were "prolonging the sufferings of the noble beast," we might just hear a little concern that he may have overstepped his rules, but we might also wonder why, if he had any concerns, he put the story into his memoir in the first place. When people write hunting memoirs, they chose the stories they want to tell and tell them the way they do for particular reasons.

In this case, one of the reasons for Gordon-Cumming chose to relate this story was that he wanted to make clear that shooting African elephants in the skull was simply not effective. Throughout the nineteenth century, hunters debated the best shots to bring down elephants, and while some hunters insisted on a shot to the skull—either just behind the ear, between the ear and the eye, or at the base of the trunk—Gordon-Cumming was convinced that it was best to aim for the side of the animal, behind the shoulder, in the hope that one might hit the heart. But even if one missed the heart, the bullet might either hit the shoulder and lame the animal or hit the lungs, which would likely lead to a strong blood spoor and slow the animal down for another shot.

But if part of the reason he included this story was to make his point about skull shots, a larger reason seems to have been that Gordon-Cumming felt the story said something important about elephants and the extraordinary experience of hunting them. Throughout his memoir, in fact, Gordon-Cumming takes pains to make clear that killing an elephant was unlike killing any other animal. And that was because, first, elephants were astonishingly powerful creatures. Describing the animals' impact on the forests, for example, Gordon-Cumming writes in an exaggerated fashion: "Here the trees were large and handsome, but not strong enough to resist the inconceivable strength of the mighty monarchs of these forests. Almost every tree had half its branches broken short by them, and at every hundred yards I came upon entire trees, and these the largest in the forest, uprooted clean out of the ground, or broken short across their stems. I observed several large trees placed in an inverted position, having their roots uppermost in the air."[26] Second, hunting elephants was a unique undertaking for Gordon-Cumming because he saw them as intelligent and secretive. Although Harris claims to have observed an almost endless valley filled with the animals, Gordon-Cumming, despite killing over a hundred elephants in the five years he was in Africa, declares that "it is only occasionally, and with inconceivable toil and hardship, that the eye of the hunter is cheered by the sight of one," adding that "owing to habits peculiar to himself, the elephant is more inaccessible, and much more rarely seen, than any other game quadruped, excepting certain rare antelopes."[27] In sum, for Gordon-Cumming "the appearance of the wild elephant is inconceivably majestic and imposing. His gigantic height and colossal bulk, so greatly surpassing all other quadrupeds, combined with his sagacious disposition and peculiar habits, impart to him an interest in the eyes of the hunter which no other animal can call forth."[28]

The regal, formidable, intelligent, and almost impossible-to-kill male elephant makes his first appearance in Gordon-Cumming's memoir near the end of the first volume. Gordon-Cumming, hunting on horseback again with a pack of dogs, finally encountered a long-sought "herd of mighty bull elephants, packed together beneath a shady grove." Within moments he had identified the most powerful of the animals, calling him "the patriarch," and naturally decided he was his target. "Cantering alongside, I was about to fire," Gordon-Cumming explains, when the bull "instantly turned, and, uttering a trumpet so strong and shrill that the earth seemed to vibrate beneath my feet, he charged furiously after me for several hundred yards in a direct line, not altering his course in the slightest degree for the trees of the forest, which he snapped and overthrew like reeds in his headlong career." The elephant pulled up and Gordon-Cumming fired at his shoulder. The elephant did not seem to notice the shot and "made off at a free majestic walk." The dogs then arrived and the elephant charged again, trumpeting as before. Gordon-Cumming shot a second time at the shoulder, but the elephant did not seem to notice. The hunter now approached the elephant on foot and shot him in the side of the head. The elephant charged again: "I stood coolly in his path until he was within fifteen paces of me, and let drive at the hollow of his forehead, in the vain expectation that by so doing I should end his career. The shot only served to increase his fury . . . and, continuing in his charge with incredible quickness and impetuosity, he all but terminated my elephant-hunting forever."[29]

Soon the elephant, with blood streaming from his wounds, was fleeing through the forest with Gordon-Cumming giving chase on his nervous horse. After a long pursuit, Gordon-Cumming was awarded with another shot at the shoulder, which was followed by another charge. The hunter followed up with another six shots from the saddle before getting off his horse to get a closer approach: "After two more shots to the side of the head, the elephant's charge came more slowly." Two more shots and the elephant charged one last time. Then two more shots to the forehead. "On receiving these," the hunter recalls, "he tossed his trunk up and down, and by various sounds and motions, most gratifying to the hungry natives, evinced that his demise was near." Another shot behind the shoulder and "the mighty old monarch of the forest needed no more." "My feelings at this moment, Gordon-Cumming concludes, "can only be understood by a few brother Nimrods who have had the good fortune to enjoy a similar encounter. I never felt so gratified on any former occasion as I did then."[30]

Nineteen shots over the course of four pages to bring down the "mighty old monarch." Part of the reason that Gordon-Cumming might have needed so many shots to kill elephants was because the musket loaders he was using were not as powerful as the large caliber guns that would be developed later in the century. However, it is not certain he did shoot any elephant nineteen times. He had a reputation for telling "large stories," so even if it took him one shot to kill this elephant or even if the events he describes never happened, his understanding of both what an elephant was and what his audience expected in a story about killing one certainly influenced this account.[31] Gordon-Cumming sought to write about killing the most powerful and noble animal in existence, a majestic creature, a monarch, and so, in his view, a mere gun should not be up to the task. It was not technology that was needed but skill and courage, along with uncommon physical, mental, and emotional stamina. His clothes ripped to shreds and night approaching, Gordon-Cumming made himself a bed of grasses by the fire and enjoyed "a piece of flesh from the temple of the elephant," which he cooked in the fire.[32]

Naturally Savage

Gordon-Cumming would likely have been surprised by Brehm's response to his memoir, but Brehm was not alone in condemning the work. Perhaps the most damning early critic, in fact, was not Brehm writing in German but James Tennent writing in English. In his *Ceylon,* Tennent devotes a chapter to elephant shooting, making it immediately clear that he finds the "sport" almost incomprehensible.[33] While he could grant that the hunt necessitated endurance on the part of the hunter, he felt that shooting elephants required "the smallest possible skill" of the marksman.[34] A hunter with a reasonable aim, he argued, should be able to kill an elephant with a single shot to the brain, and he points to hunters who had accounted for hundreds of slain elephants with such shots.[35] Although he found the "wholesale slaughters" exacted by the hunters in Ceylon to be a "monotonous recurrence of scenes of blood and suffering," he did praise them for the efficiency of their kills, to which he contrasts the methods of Gordon-Cumming's hunts, decrying the "sickening details" that he provides of the "torture inflicted by the shower of bullets which tear up" the animals' flesh.[36] To make his point, Tennent quotes at length two of Gordon-Cumming's descriptions that he felt were particularly egregious—the long description of the hunt that included the testing of various shots at the elephant's head and a hunt in which Gordon-Cumming claimed to have fired forty times leaving the "elephant's forequarter" "a mass of gore" and the animal "trembling violently beside a thorn tree," where he "kept pouring water into his bloody mouth until he died."[37] Calling Gordon-Cumming's hunts "motiveless massacres" that lacked any "manly justification," he suggests he could regard "almost with favor" the efforts of Thomas

William Rogers, who allegedly applied the money he gained from the ivory of killing over fourteen hundred elephants to various regimental commissions and thereby at least had an objective in his hunting.[38]

While Tennent finds Gordon-Cumming's behavior contemptible, he recognized that he was hardly unique among elephant hunters. In Ceylon, Tennent found Gordon-Cumming's contemptable equal, it seems, in the British hunter Samuel White Baker, who published an account of his hunting adventures in Ceylon in 1854 under the title *The Rifle and the Hound in Ceylon*. But if Tennent saw Baker and Gordon-Cumming as of a piece, other observers regarded the two men and their hunts very differently. In an 1854 article, *Harper's New Monthly Magazine*, for example, categorically states that Baker was "a hunter of quite a different stamp" from Gordon-Cumming. Describing Gordon-Cumming's hunting as "little better than butchery," the editors insist that the pleasure in the hunt for Baker arose "not as much from the death of the animal, as from the skill and courage demanded on the part of the hunter."[39]

Baker also seems to encounter a very different creature from the object of Gordon-Cumming's hunts. Baker begins his account of what he calls the most "misunderstood" of animals by describing the elephant in a way that both builds on and diverges from Gordon-Cumming's portrayal: "Lord of all created animals in might and sagacity, the elephant roams through his native forests. He browses upon the lofty branches, upturns young trees from sheer malice, and from plain to forest he stalks majestically at break of day, 'monarch of all he surveys.'"[40] Gordon-Cumming would never have used the word "malice" to describe the motives of an elephant, but otherwise the description echoes that of the South African hunter. As Baker continues, however, his distance from Gordon-Cumming's mindset becomes more clear. Calling the elephants seen in zoological gardens "unwieldy and sleepy looking" and apparently interested only in catching bits of food thrown by children, he insists that few people have direct experience with the true, wild elephant. The "sire" of the pathetic captives, Baker writes, may well have been "the terror of a district, a pitiless highwayman, whose soul thirsted for blood; who, lying in wait in some thick bush, would rush upon the unwary passer-by, and knew no pleasure greater than the act of crushing his victim to a shapeless mass beneath his feet."[41] Wild elephants, according to Baker, are "naturally savage, wary, and revengeful, displaying as great courage when in their wild state as any animal known," and their "great natural sagacity" makes them even "more dangerous as foes."[42]

Among all elephants, Baker continues, the most dangerous are the solitary males, the "rogues," who choose to leave the society of other elephants and subsequently become fantastically vicious beasts who terrorize whole areas of forest. Claiming that they always walk downwind so that they will be able to scent any pursuer, Baker states that they will freeze motionless—"like a statue in ebony"—when they become aware of a hunter. At that moment, the hunter becomes the hunted. One time, he relates, he and his brother were out when they realized too late that they had been craftily led into a trap—a small muddy clearing surrounded by impenetrable jungle—by two rogues who had apparently teamed up in order to wreak more havoc. Baker's brother got stuck in the mud, and then they "suddenly heard a deep guttural sound in the thick rattan." "In the same instant," Baker writes, "the whole tangled fabric bent over me, and bursting asunder showed the furious head of an elephant with uplifted trunk in full charge upon me" (fig. 3.2).[43]

Figure 3.2. Samuel White Baker, "Narrow Escape," in *The Rifle and the Hound in Ceylon* (London: Longman, Brown, Green, and Longmans, 1854). American Geographical Society Library, University of Wisconsin–Milwaukee Libraries.

The brothers both fired. Then Baker jumped to avoid the charge, but his feet got tangled in the grasses and he fell to the ground directly in front of the charging animal. Expecting to hear the "crack" of his own bones, Baker heard instead the "crack of a gun" as his brother fired again. Feeling a "spongy weight" against his heel, Baker rolled out of the way and saw the downed elephant still trying to reach him with his trunk. Grabbing his "four-ounce rifle" from a gun bearer, he was preparing to give the elephant "a finisher" in the head when the other elephant then attacked with a "savage scream." Baker writes, "I saw the ponderous fore-leg cleave its way through the jungle directly upon me. I threw my whole weight back against the thick rattans to avoid him, and the next moment his foot was planted within an inch of mine." Having only his smaller rifle and realizing his brother's defenselessness, Baker fired "perpendicularly" at the animal's throat as it passed over him, jumping away "the instant after firing." The elephant stopped in its charge against Baker's brother, staggered off with a severed jugular pouring blood, and was discovered dead several days later.[44]

If Tennent thought the gore of Gordon-Cumming's stories was disgusting, he found Baker's accounts simply absurd. Referring to the "tiresome iteration" of hunting stories that seemed designed only to glorify the hunter, Tennent argues that while it may be true that rogues were dangerous, they were also so rare—at most only one in five hundred elephants was a rogue, in his view—that few hunters would ever likely encounter one. Indeed, the proposition that elephants would be "thirsting for blood" while "lying in wait in the jungle" was, Tennent maintains, simply ridiculous. If anything, he insists, "the cruelties practised by the hunters have no doubt taught these sagacious creatures to be cautious and alert, but their precautions are simply defensive; and beyond the alarm and apprehension which they evince on the approach of man, they exhibit no indication of hostility or thirst for blood."[45]

It is easy to see why Tennent would find Baker's stories about "rogues" ludicrous. At one point, for example, after Baker and his brother had killed nine elephants in a matter of moments—underscoring, again, that it might not be so difficult to kill an elephant—Baker claims he was attacked by a rogue: "From the very spot where the last dead elephant lay, came the very essence and incarnation of a 'rogue' elephant in full charge. His trunk was thrown high in the air, his ears were cocked, his tail stood high above his back as stiff as a poker, and screaming exactly like the whistle of a railway engine, he rushed upon me through the high grass with a velocity that was perfectly wonderful. His eyes flashed as he came on, and he had singled me out as his victim."[46]

Baker decided his only hope was to stand his ground and hope he could kill the elephant with the one barrel he had left with a shot to the animal's head and then somehow leap to the side at the last moment. "There could not be a better exemplification of a rogue than in this case," he states. "A short distance apart from the herd he had concealed himself in the jungle, from which position he had witnessed the destruction of his mates. He had not stirred a foot until he saw us totally unprepared, when he instantly seized the opportunity and dashed out upon me."[47] Like the story of the two rogues who worked together, one of the basic problems with this story is that Baker's definition of a "rogue" is a male elephant who lives a solitary life, becomes morose, and then, motivated by an evil impulse, attacks and destroys whatever comes into its path. In both of these stories, what makes the animals exemplifications of rogue is not that they lived alone, not that they attacked without cause, but that they were male elephants and that they charged after the shooting had begun, a behavior Gordon-Cumming, for example, would have thought to have little to do with malevolent intent.

Despite Tennent's efforts, with the stories of Baker and others, the idea of the terrifying, crazed, and vicious elephant became increasingly popular among hunters. Before Baker's characterization of elephants gained traction, an elephant could seek justified revenge against a personal or perhaps familial attack, but that response was tempered by the animal's modesty and rationality; after, wild adult male (and even sometimes female) elephants could essentially always be described as "rogues." Labeling an elephant as a "rogue," of course, would then provide a ready excuse to kill it. If permits were increasingly needed to shoot elephants, a hunter was always justified in killing an attacking elephant, and many so-called rogues in the period were clearly just elephants with ivory or elephants whose tails were needed as proof of the hunter's prowess. With that said, the emergence of the idea of the rogue in the middle of the nineteenth century seems to have been part of a larger shift in ideas about elephants and "dangerous" animals more generally. As the century wore on, animals like elephants, lions, tigers, sometimes gorillas, and a few others became increasingly interesting to hunters because they could, in their view, be *both* spectacularly dangerous and chillingly cunning. Earlier hunters had seen certain animals—especially elephants—as difficult to hunt and kill, but these hunters never seem to have conceived of their quarry as guileful, as driven by bloodlust, or as just plain man-killers. Those ideas would become more and more popular in the second half of the century, a time when readers were more likely to live in cities and not themselves be hunters. If Edwin

Landseer's *Monarch of the Glen* captured many of the qualities of the elephant of Buffon, Harris, and Gordon-Cumming, Eugène Delacroix's contemporary paintings of lions and tigers attacking powerful and noble horses capture better the blood-thirsty elephants of Baker.

A Creature of the Pleistocene

It was in the context of such new ideas about elephants that Theodore Roosevelt, just weeks after the end of his presidency in March 1909, embarked on an eleven-month safari to East and Central Africa, known as the Smithsonian-Roosevelt African Expedition. Officially, the trip was organized to secure needed specimens for the new natural history building of the Smithsonian, which would open in 1910. In the end, the expedition, which included the former president and his son Kermit, scientists from the Smithsonian, including the ornithologist Edgar Mearns and zoologists Edmund Heller and John Loring, and the guide R. J. Cuninghame, secured over eleven thousand specimens for the museum. Theodore Roosevelt logged 296 animals in his official register; Kermit, 219. The trip was financed through three sources: Andrew Carnegie, funds donated to the museum, and Roosevelt, himself. Roosevelt's contribution came from money he received through a contract with *Scribner's Magazine* to write a serialized report of the trip, the first installment of which appeared in October 1909 and the last in September 1910. The installments were then collected and published in 1910 as *African Game Trails: An Account of the African Wanderings of an American Hunter-Naturalist*. In the foreword to the book, now one of the classics of hunting literature, Roosevelt writes in tones that echo Baker. Beginning with a line from Shakespeare's *Henry IV, Part II*, that seems remarkably fitting for a person who had just left the presidency, from a moment in the play when the fat knight Falstaff hears from his ensign news of the death of the king, news portending changes, adventures, and happy prospects to come, Roosevelt writes: "'I speak of Africa and golden joys'; the joy of wandering through lonely lands; the joy of hunting the mighty and terrible lords of the wilderness, the cunning, the wary, and the grim."[48]

There are many, many hunting stories in *African Game Trails*; indeed, Roosevelt and his party started hunting just days after they arrived and were essentially at it with only brief breaks for the whole trip. At the end of the text Roosevelt provides a list of the game the party shot using a rifle that begins with the lion, leopard, cheetah, hyena, elephant, square-mouthed rhinoceros, hooked-lipped rhinoceros, hippopotamus, wart-hog, common zebra, big or Grevy's zebra, giraffe, buffalo, giant eland, common eland, bongo, kudu, sitatunga, East African bushbuck, Uganda harnessed bushbuck, Nile harnessed bushbuck, sable, roan, oryx, wildebeest, Neumann's hartebeest, Coke's hartebeest, Jackson's hartebeest, Uganda hartebeest, Nilotic hartebeest, and topi and that goes on to list another fifty species. To this, Roosevelt acknowledges, one would have to append the birds collected with shotguns and the animals shot to feed the large expedition are likewise not mentioned. In addition, the expedition was also collecting plants, fish, invertebrates, live animals, and ethnographic artifacts. That the official list includes thirty-eight species of antelope ordered generally from larger and more rare to smaller and more common and that it begins with lion, leopard, cheetah, hyena, elephant makes it clear that whatever its scientific goals and outcomes, this trip was one of the largest hunting holidays ever mounted. And that the former president keeps track only of his shots at large game while the scientists collect everything else, that he gets the first shot, that his "bag" is the largest, that he sits down to enjoy one of his books as the scientific team begins the work of skinning and preserving the specimens, and that a large American flag always flew before his tent also demonstrates that the trip was a kind of political theater mimicking the hunts or European aristocracy. Still, although the trip was, above all else, about the glorification of the former president and his hunting passions, it had other less egocentric goals, and Roosevelt's interest in natural history was sincere and long-lived.[49]

By the time of the trip, Roosevelt had had many hunting adventures in North America, but Africa seems to have been altogether new for him. In his opening chapter, "A Railroad through the Pleistocene," he describes the journey from Mombasa to Lake Victoria at the beginning of his adventure as a trip through time. The idea of travelers moving through time would have been familiar to audiences who had read such works as H. G. Wells' 1895 *The Time Machine* and Jules Verne's 1864 *Journey to the Center of the Earth*, as would have been the idea of an animal or human returning to a primeval mode of life to those who had read Jack London's 1903 *Call of the Wild* or Rudyard Kipling's 1894 *Jungle Book*. Roosevelt describes how the railroad on which he traveled from the coast to the interior "pushed through a region in which nature, both as regards wild man and wild beast, did not and does not differ materially from what it was in Europe in the late Pleistocene Age." With its endless herds, giant and terrifying creatures, and "savage tribes," the Africa he saw reproduced "the conditions of life in Europe as it was led by our ancestors ages before the dawn of anything that could be called civilization." "African man," he insists, "absolutely naked, and armed as our early paleolithic ancestors were armed, lives among, and on, and in constant dread of, these beasts, just as was true of the men to whom the cave lion was a nightmare of terror, and the mammoth and the woolly rhinoceros possible but most formidable prey."[50] Roosevelt was using a popular idea and an almost conventional way of writing about "primitive" people and the places they lived. When Africans and other "exotic" peoples were exhibited in Europe and the United States in the popular ethnographic displays of the time, for example, newspapers would often report that visiting the shows was like traveling back through time. Nevertheless, Roosevelt's extended reflections on this topic suggest, as he insisted, that the idea was more than just "fanciful." Recalling his experiences in Africa years later in an essay entitled "Primeval Man; and the Horse, the Lion, and the Elephant," for example, Roosevelt writes: "Day after day I watched the thronging herds of wild creatures and the sly, furtive human life of the wilderness. Often and often, as I so watched, my thoughts went back through measureless time to the ages when the western lands, where my people now dwell, and the northern lands of the eastern world, where their remote forefathers once dwelt, were filled with just such a wild life."[51] "In that immemorial past," he continues, "the beasts conditioned the lives of men, as they conditioned the lives of one another; for the chief factors in man's existence were then the living things upon which he preyed and the fearsome creatures which sometimes made prey of him."[52]

It is not that Roosevelt thought that Africa was locked in the deep past, but he clearly felt that the lives of the animals and peoples of Africa were structured by an ancient struggle for existence, a struggle dominated by hunting and being hunted. Even though the people with whom he came into contact were typically agriculturalists and likely threatened more by disease and hunger than lions and elephants, the focus of Roosevelt's account—as is signaled by the frontispiece trophy shot of him posing, gun in hand, above a male lion he had just killed (fig. 3.3)—remains an Africa filled with deadly beasts.[53] In recalling his ancestors, Roosevelt suggests that as much as the expedition served scientific purposes, it was also about connecting to what he considered essential elements of human existence, about casting aside the comforts of civilized life and testing himself against ancient foes. The fact that these tests came with such powerful weapons as a Holland and Holland double rifle firing a .500 / 450 Nitro Express cartridge and a Winchester .405 was apparently immaterial. It seems that Roosevelt hankered for a sort of primal experience that he believed could be procured by standing his ground, gun raised, before

Figure 3.3. Theodore Roosevelt with a dead lion, frontispiece to *African Game Trails* (New York: Charles Scribner's Sons, 1910). Photograph by Kermit Roosevelt.

a charging lion, buffalo, rhinoceros, or elephant. Years later he recalled an evening he spent, apparently alone, at the site of a recent lion kill, and his walk back to his camp through a gorge, "an eerie place" as "darkness fell." Aware of the dangers around him, his "thoughts went back through the immeasurable ages to a past that was always dangerous; to the days when our hairy and low-browed forefathers, under northern skies, fingered their stone-headed axes as they lay among the rocks in just such a ravine as that I had quitted, and gazed with mingled greed and terror as the cave-lion struck down his prey and scattered the herds of wild horses for whose flesh they themselves hungered."[54] Roosevelt did not simply deploy these stories of prehistoric pasts for literary effect; they are, rather, critical to understanding his safari to Africa.

Roosevelt began the hunt for his first elephant on the slopes of Mount Kenya, accompanied by Cuninghame, two guides, his gun bearers, and some porters. He writes, "we struck into the great forest, and in an instant the sun was shut from sight by the thick screen of wet foliage. It was a riot of twisted vines, interlacing the trees and bushes. Only the elephant paths, which, of every age, crossed and recrossed it hither and thither, made it passable."[55] Roosevelt uses his description of the forest—foreboding, dangerous, and unyielding—to set the tone for this hunt. This will not be a case of a wild ride behind a pack of dogs chasing a mountain lion in the American West; here, Roosevelt pits himself against nature itself. As he continues to describe the forest, he notes stinging plants, mosses and ferns "rank and close," and trees that were of "strange kinds." In some places the trees were lower with the ground thick with bushes and shrubs. In other places, though, "mighty monarchs of the wood, straight and tall, towered to an immense height." "Far above our heads," he writes, "their gracefully spreading branches were hung with vines like mistletoe and draped with Spanish moss" and their "buttressed trunks were four times a man's length across."[56]

After a long and tense stalk on a small herd of elephants, Cuninghame and Roosevelt spotted "the grey and massive head of an elephant resting his tusks on the branches of a young tree." Determining it was a bull with "good ivory," Roosevelt fired. Although he hit the animal "exactly" where he aimed, the shot only "momentarily stunned the beast." The elephant "stumbled forward," and as he pulled himself up, Roosevelt fired his "second barrel, again aiming for the brain." As he lowered his gun, he "saw the great lord of the forest come crashing to the ground."[57] The story does not end there, though. Echoing Baker's account of his hunt of the two "rogues" with his brother, Roosevelt adds that at "that very instant," before there was time to reload, the "vast bulk of a charging bull elephant" came at him from his left, crashing through the undergrowth. "He was so close," Roosevelt explains, "that he could have touched me with his trunk." Not surprisingly, I suppose, Roosevelt describes himself jumping away at the last moment, "opening the rifle, throwing out the empty shells, and slipping in two cartridges." At the same moment, his hunting companion Cuninghame fired both his barrels and threw himself into bushes as well. Cuninghame's shots apparently stopped the charge, and the elephant turned and quickly vanished into the forest. "We ran forward," Roosevelt concludes, "but the forest had closed over his wake. We heard him trumpet shrilly, and then all sounds ceased."[58]

Roosevelt notes that had they only been in Africa for ivory, they would have followed the second bull, but "there was no telling how long a chase he might lead us; and as we desired to save the skin of the dead elephant entire, there was no time whatever to spare." So, no following up on a wounded animal in this case—science must come first. The party returned to "where the dead tusker lay" to begin the "formidable task" of preserving the skin.[59] Before skinning the animal, though, Roosevelt organized a series of photographs of different members of the hunting party standing with the dead bull. The photographs—one, for example, by Edmund Heller of Roosevelt leaning on the dead elephant (fig. 3.4) with his African guides and carriers barely indistinguishable from the forest—make clear that it was deeply important to Roosevelt to record in every way he could his first elephant hunt. The skull, bones, skin, tusks, and photographs of this elephant were above all trophies marking Roosevelt's success as a hunter, not his vaunted role as a scientific

Figure 3.4. Roosevelt's first bull. Photograph by Edmund Heller, Smithsonian Libraries.

naturalist. The importance of ignoring the distraction of the ivory as well as the ethics of pursuing the wounded elephant and of getting started with the skinning and preserving without delay notwithstanding, there was apparently time enough to haul the camera to the site, set it up, clear all the trees, vines, and other vegetation obstructing the view from the camera to the fallen elephant (this task alone would have taken ages), and take photographs of the great accomplishment. There was more than enough time, then, before the scientific work began for Roosevelt to savor a triumph that, he wants us to believe, had absolutely nothing to do with being in Africa to pit himself against the most dangerous game and be certain that that game carried as much ivory as possible. As Roosevelt himself concludes: "I felt proud indeed as I stood by the immense bulk of the slain monster and put my hand on the ivory."[60]

A month after the bull hunt on Mount Kenya, the Roosevelts, along with Cuninghame and one of the outfitters for the safari, Leslie Tarlton, were hunting elephants again. This time they were east of Mount Kenya in the area known as Meru. Roosevelt begins by describing the difficulty of seeing the animals in the dense forest and recounts how Tarlton stood on Cuninghame's shoulders to get a better view. Eventually, the party, standing precariously on a six-foot diameter trunk, noticed a tree shaking in the forest. The hunters carefully stalked the group of elephants, and once they were within range, Roosevelt fired his gun at the largest elephant of the group, stunning it, and then he fired his second barrel, which knocked it down. Roosevelt then turned to another elephant, firing two more shots. At that moment, he realized his first elephant was up, so he fired another two shots at it, knocking it down again. After it rose a second time, he fired two more rounds. His seventh shot, which he took with a smaller rifle, finally finished off the first elephant, and Roosevelt took off after the fleeing and wounded second elephant.

At one point in the description of the hunt, as the party was quietly approaching the animals, Roosevelt writes: "We could not tell at what second we might catch our first glimpse at very close quarters of 'the beast that hath between his eyes the serpent for a hand.'"[61] Roosevelt was a constant reader, and among the volumes he carried with him on his trip to Africa in his specially prepared "Pigskin Library" were histories, essays, and poems by one of his favorite authors, Thomas Babington Macaulay, the British historian, essayist, poet, politician, and much more.[62] In *The Prophecy of Capys*, one of his immensely popular "Lays of Rome," Macaulay describes Pyrrhus of Epirus's attack on Rome in the third century BCE. The blind seer Capys warns Romulus:

The Greek shall come against thee,
The conqueror of the East.
Beside him stalks to battle
The huge earth-shaking beast,
The beast on whom the castle
With all its guards doth stand,
The beast who hath between his eyes
The serpent for a hand.[63]

Recalling heroic battles of the past, the lines from Macaulay seem well chosen. Many of Roosevelt's readers—especially the educated readers he seems to have imagined as his audience—would have recognized the poem and would perhaps have enjoyed the description of Roosevelt's apparently epic hunts with such an epic reference. This was the beast that Roosevelt sought and killed.

The Last Fell with Its Head Resting on the Other

"The chase of the elephant, if persistently followed," Roosevelt opines, "entails more fatigue and hardship than any other kind of African hunting." Roosevelt admitted that he suspected that hunting lions was more dangerous, but he insists that "far greater demands are made by elephant-hunting."[64] Part of what made Roosevelt's elephant hunting difficult was that he was pursuing them in thick forest, but even with that caveat it should be clear that for hunters like Roosevelt shooting elephants represented a kind of ultimate test. To stand before an animal that "shakes the earth" and kill it said more about the hunter's moral strength than the power of the weapon he held in his hands. However, as much as Roosevelt and his hunting friends wanted readers and listeners to appreciate what they perceived as the grandeur of shooting elephants, of standing beside fallen monsters with a hand resting on the ivory, there were other hunters whose accounts made it clear that shooting elephants might be little more than a kind of mechanical slaughter. Among these was Arthur Henry Neumann, whose *Elephant Hunting in East Equatorial Africa* was published in 1898, more than a decade before Roosevelt's *African Game Trails*.[65]

Unlike Roosevelt, Neumann was not well known before he published his memoir. But the exceptionally well-produced volume, illustrated by known artists, including John Guille Millais, the son of John Everett Millais, was admired for what seemed its believable if not particularly vivid accounts. In the 1890s, after having lived in Africa for over twenty years, Neumann decided to become a commercial elephant hunter, collecting his own tusks and purchasing tusks from others. In Neumann's account, although there are a few hair-breadth escapes, there is little contemplation of the magnificence of the hunt, something central to the memoirs of Gordon-Cumming, Baker, Roosevelt, and so many others. Neumann's descriptions were not necessarily shorter, but they focused more on logistics and techniques and spent little time describing the animals or dwelling on the hunter's deep emotions.

In his account of a particular red-letter day, for example, Neumann begins by noting that he was traveling with his party when "elephants were seen in the bush below." He continues:

> I at once commenced the campaign, getting first to leeward and then cautiously advancing. I easily got close up, and could see several, as they were standing in a most favourable position by trees where there was a small space comparatively open, while I was hidden by a screen of tall undergrowth; but as one was much larger than the rest (evidently a big bull, though I could not see his tusks), I waited till he gave me a chance at the temple, and was lucky enough to drop him dead, killing a cow alongside of him similarly with the second barrel. Loading quickly I had time to knock over another cow before they ran.[66]

A few sentences and three dead elephants. From here, Neumann pauses to describe the efficacy of well-placed head shots and explains that if one can drop an elephant in one shot, it usually takes a few moments before the other elephants decide what to do, thus giving the hunter an opportunity to kill more of the animals. Neumann adds that he wasted no time admiring the animals he had killed because he heard others close by. He quickly climbed a tree and was able to make one out. Stalking that elephant, Neumann then heard something to his left. He turned "to behold another, which must have been within a few yards, though hidden by the thick high scrub, coming for me and almost on me." He did not have time to aim, he writes, so he shot at the head and jumped. The female elephant was not stopped and stood "screaming and wondering, as it seemed, what had become of me." In an awkward position and "a bit flustered, too," Neumann fired his second barrel, and the elephant made off.[67]

After finding the cartridge pouch he had temporary misplaced, he takes off after the cow he had wounded, but then he notices another clump of elephants: "Getting up to them, I succeeded in flooring another cow with fine tusks, one of which grew right across the other (this skew tooth proved particularly long and solid, having scarcely any hollow at the base). The rest ran to the edge of the bush and stood on a slight rise among low scrub, just outside the tall forest patch. I followed, and getting a good view of them knocked over a right and left, and loading again was just about to repeat the performance, as the others still stood about apparently quite bewildered, when I was suddenly attacked by swarms of bees."[68] Neumann pauses to describe the practice of hanging "tubs" in trees to attract wild bees and then returns to his hunt, wounding another cow before he "felled her again" with his second barrel and "left her for dead."[69] Receiving news that a herd had been seen on the other side of the valley, Neumann headed in that direction. "Before emerging from the forest," though, he "came across another little lot, and, my luck still holding, I scored another right and left—the third to-day—at a young bull and a cow."[70] Continuing on, he shot at a larger elephant "but bungled it somehow" and then, noticing some "natives" beckoning, he found another small herd. Facing difficulty getting close for a clear shot, Neumann climbed up a tree and got a good view, where he waited for the "largest to give me a chance."[71] Deciding to shoot another, he writes, "I chose a cow with long and very straight tusks, which stood at the outside of the clump, giving me a fine chance at her temple, and had the satisfaction of seeing her throw up her trunk and fall dead to my shot." Having used up all the cartridges for his large gun, he grabbed his single-shot Martini-Henry. The elephants were moving away but still within shot. He saw two "with their heads in view long enough" to give him another chance, and he killed both. "The last fell," he wrote, "with its head resting on the other. It proved to be a young one with very small teeth, unfortunately, but in such thick jungle it is impossible to pick and choose much. I had now eleven elephants dead."[72]

Neumann was "well pleased" with the day, but he was disappointed he had not killed more, "particularly the big fellow I had wounded."[73] The large elephant was "found some time later," though, and he recovered its tusks, which "weighed between 80 and 90 lbs. apiece."[74] In six pages and twenty shots, Neumann had killed twelve elephants. The tails were cut off immediately to prove his ownership, and the next day, the hunter moved his camp to the area where he had been so successful so his men could start removing the ivory. Arranging his new camp by a stream, he was able to shoot a giraffe as well, so he could conveniently eat more tasty giraffe over the next days instead of elephant.[75]

Visiting Neumann's old boma in Meru, Roosevelt described his predecessor as "once the most famous elephant-hunter between the Tana and Lake Rudolf," a "mighty hunter, of bold and adventure-loving temper."[76] He also admired Baker—"a mighty hunter and good observer"—and found himself wistful about the times of Harris and Gordon-Cumming.[77] On days when nothing special was planned, Roosevelt wrote that he liked to go riding with only the company of his servant and gun bearers. "I cannot describe the beauty and the unceasing interest of these rides, through the teeming herds of game," he states. "It was like retracing the steps of time for sixty or seventy years, and being back in the days of Cornwallis Harris and Gordon Cumming, in the palmy times of the giant fauna of South Africa."[78] Harris, Gordon-Cumming, and Baker were names, along with those of Selous, Stigand, Patterson, Andersson, Powell-Cotton, Sanderson, and others, synonymous with legendary hunting adventures for Roosevelt and his readers.[79] As Roosevelt's party was making its way to Africa, steaming across the "hot, smooth waters of the Red Sea and the Indian Ocean," they listened to the stories of Frederick Courteney Selous, who was also on board and whom Roosevelt declared "the greatest of the world's big-game hunters." For his rapt listeners, according to Roosevelt, Selous conjured "strange adventures that only come to the man who has lived long the lonely life of the wilderness."[80]

For Roosevelt, and I think for many of other big-game hunters of the time, there seemed a sort of assumed and natural camaraderie among fellow hunters, a sense that they all had similar experiences and similar feelings about hunting. When Gordon-Cumming styled himself as a "chief" over all he surveyed and described his feelings of triumph when he inspected the elephants he killed, I think he was defining a feeling that many hunters who have proudly posed with a "trophy" would recognize. It may be true, as the philosopher Mary Midgely insists, that Gordon-Cumming was, in fact, only a man exhibiting "glaring faults of confused vainglory and self-deception," but his sentiments have nevertheless been expressed over and over again in hunting memoirs.[81] Still, while hunters like Harris, Baker, and Roosevelt might easily be grouped together because they shared an idea about hunting as a certain kind of thrilling experience, it is clear that their sensibilities about what they were doing, about what shooting an elephant might represent, were still quite different from one another. In the end, not all hunters understand the "game" and their participation in it in the same ways. However much, for example, Roosevelt may have admired people like Selous, he was often regarded with restrained enthusiasm or simple contempt by other hunters.[82] In the end, all these hunters—Gordon-Cumming, Harris, Neumann, and Roosevelt, and even Brehm, because he too was an avid hunter—had different ideas about hunting, about hunters, and about elephants. Together, though, their memoirs—the stories they told about hunting—helped define new ways of thinking about this remarkable animal.

Before the middle of the nineteenth century, Harris and Gordon-Cumming, hunting in southern Africa, pitted themselves against a monarch of the forest. These men were among the early African sport hunters, those new figures in the landscape who hunted for adventure and not simply for meat or as part of settling the land. They portrayed themselves as inheritors of traditions of aristocratic hunting and vividly described their adventures to their readers and those who attended their lectures in elevated terms as focused before anything else on shooting "noble" game. Yes, they killed almost anything they could find to feed themselves and their parties, and they collected and sold ivory, but when they described "a hunt," they always depicted themselves seeking something more than food or white gold. In their stories, the elephant—the creature described before them best by Buffon—was an animal of immense power and intelligence and killing it was a task beyond the abilities of most men. From the middle of the nineteenth century to the beginning of World War I, however, a new image of the elephant and of elephant hunting emerged. In works by Baker and Roosevelt and the many, many like them hunting in India, Ceylon, and East and Central Africa, the elephant would become a vicious and deceitful rogue that was as much predator as prey, and the forest and jungle would become foreboding places where only men of exceptional courage and stamina could succeed. As James Sutherland reveled in his 1912 memoir *The Adventures of an Elephant Hunter*, "I think it would be difficult to find another [life] so full of wild, exhilarating excitement, hair-breadth escapes, and devil-may-care risks, and though the end is usually swift, perhaps that is better than flickering out slowly on a bed of sickness."[83]

When we look back on elephant hunting in the nineteenth century, it is all too easy to imagine a sort a stereotype, perhaps even buffoon of the nineteenth-century big-game hunter—a "little man" (and it is almost always a man even though there were women who sometimes hunted with them) dressed in khaki with a pith helmet, a man who proclaimed his power through his assertion of "superior" culture and who exercised that power with increasingly formidable guns.[84] George Orwell touched a deep truth about people like Baker and Roosevelt, I think, when he wrote that "every white man's life in the East, was one long struggle not to be laughed at."[85] The ridicule and scorn, however, came not only from the indigenous people who had to carry the champagne, the food, the tents, the trophies, and even the hunters who didn't want to get wet crossing rivers. While people at home undoubtedly spoke under their breath about the "great" hunters, others, like Brehm and Tennent, made their opinions clear in print.[86] In the United States, voices like that of Reverend William J. Long—who had sparred many times with Roosevelt over the years—said what many surely thought when he was quoted in the *New York Times* calling the former president "a game butcher pure and simple" whose "interest in animals lies chiefly in the direction of blood and brutality."[87]

Whether seeking pure sport, confronting an ancient foe, or exploiting a resource, all the hunters in this chapter contributed to the early history of debates that are still with us regarding both the sport hunting of elephants and the commercial harvest of ivory. Ideas that emerged about elephants and hunting in the long nineteenth century changed as the sport evolved and adopted new technologies, as more people became familiar with elephants through visits to zoos and circuses, and as the game itself became increasingly scarce.[88] As larger debates about the advantages and disadvantages of colonial empires unfolded and stories about and justifications for the hunts began to change, the audiences back home were also becoming more familiar with actual elephants and more skeptical of the claims of the so-called great white hunters. While Harris, Gordon-Cumming, Baker, Roosevelt and Neumann were shooting elephants in the colonies, elephants were becoming an increasingly important presence back home.

The Most Friendly Creature

Lily was a little less than three months old when I first met her (fig. 4.1). It was the second of a series of visits to the Oregon Zoo I had made with a photographer friend for a project we were working on about relationships between elephants and keepers.[1] To have been able to do the project anywhere would have been an extraordinary opportunity, but to be able to do it while meeting and spending time with both one of the oldest elephants in North America—an elephant I had read about since I was a kid—and the youngest was much more than I ever expected. It was honestly difficult to find a place that was open to the project. I have typically been greeted with a healthy skepticism from zoo directors about my wanting to know more about the animals and the history of their institutions. This is not too surprising. Most people who want to go behind the scenes at zoos can be grouped easily enough. There are the members of the public who are fascinated by the animals and look forward to announcements of special "members-only" tours as opportunities to learn more and see different perspectives on the institutions they already enjoy. There are critics of zoos who hope to discover and spread awareness about what they see as unethical practices. There are scientists with easily checked credentials who want to collect samples or data or learn from the keepers. There are local journalists looking for a good story, maybe a predictable human interest piece, and there are other journalists working on deeper investigative stories. The first group of reporters show up regularly and become familiar to media relations staff, the leadership and curators, and even some of the keepers. The second group is potentially more problematic, and most of the people responsible for deciding who warrants the sort of consideration from staff members that would take them away from their normal duties would naturally lump me together with investigative journalists. It would

Figure 4.1. Lily at seventeen months, April 2014. Author's collection.

frankly be naive for a zoo professional to think that people wanting to dig deeply into a zoo's history and its practices would come without an agenda because as much as zoos may be extremely popular institutions, they have very outspoken critics as well. It seems that almost every zoo official has stories to tell about people showing up declaring their open-mindedness to the concept of zoos whose actual agendas only became clear much later on.

Most people who do not like zoos talk about the animals looking depressed or bored. They are convinced that "wild animals" should not be kept in captivity because their needs could never be met in such unnatural settings. They point to signs of psychological distress in many of the animals. They cite examples of unethical practices in the zoo and aquarium industry such as euthanizing or selling surplus animals. They point to ineffective and overhyped educational efforts and scientific studies without merit. They argue that profit motives drive the exhibition of certain especially charismatic animals like whales, polar bears, and elephants and they insist that breeding endangered species is prompted more by a desire to preserve the animal in captivity than in the wild. They are often convinced that the keepers are cruel and that the public is simply ill informed or more interested in their own pleasures than the suffering of others. To those who claim that zoos are institutions fostering science, education, conservation, and recreation, they would generally say that H. L. Mencken was right when he wrote over a century ago that a zoo is nothing more than "a childish and pointless show for the unintelligent" and suggested that "the sort of man who likes to spend his time watching a cage of monkeys chase one another, or a lion gnaw its tail, or a lizard catch flies, is precisely the sort of man whose mental weakness should be combatted at the public expense, and not fostered."[2] Zoos exist, he concluded, only so that "a horde of superintendents and keepers may be kept in easy jobs" and so that "the least intelligent minority of the population may have an idiotic show to gape at on Sunday afternoons."[3]

The many thoughtful people I have met over the years who work in zoos understand these criticisms and acknowledge that zoos often fall short of the rhetoric supporting them. The institutions, they acknowledge, inevitably have less than ideal facilities for animals because one could and even should do more and better. They know that most of the people who visit are not that interested in learning, and they know, too, that the staff at zoos—from the leadership down to the keepers—can make the wrong decision from time to time when faced with the complexities of caring for their charges. It has always seemed to me, though, that the motivations of most zoo professionals are transparent enough. People do not become animal carers at major zoos for money or prestige. They do the work because it is deeply fulfilling to them. Very few of us are lucky enough to have a job doing exactly what we had always hoped to do. All the people I have met who work directly with animals at major zoological gardens feel they are doing just that.

When I approached the director and the elephant keepers at the Oregon Zoo in Portland about coming to study the day-to-day interactions of keepers and elephants at the zoo, I was greeted with a cautious optimism and openness about where the project might lead. They knew my colleague and I were working on a book with photographs and text, but we did not promise anything other than to try to represent what we actually saw rather than the usual storylines of marketing departments and critics. We visited for multiple full days at a time over the course of a couple of years. We began to know a little about the staff, the volunteers, and the elephants. We also got to know the building. At the time, it was some sixty years old. When it was originally constructed, it incorporated all the best ideas then current about holding elephants, and it was updated periodically over the decades. There were large outdoor yards with natural substrate designed to promote the health of the elephants' feet and joints, swimming pools, relatively easy-to-clean concrete floors and walls in the building, a central "sunroom" for washing the animals daily and for the animals and staff to safely interact with each other, massive doors and gates that were hydraulically operated, a squeeze chute (the first in an elephant house in the United States) to allow the keepers and veterinary staff close and safe access to the animals, and a variety of different locations from which the public could view the animals, who were seen typically outdoors in the mild Portland, Oregon, weather. When I first visited the old building, I was struck by how much better it was for the animals and the keepers than so many buildings I had seen over the years. Yet the building was increasingly seen as inadequate—not because it couldn't meet the basic needs of the elephants but because ideas of elephant husbandry and the technologies and design concepts of the leading-edge zoos today are just so very different from those of the middle of the twentieth century. It was razed, finally, in the summer of 2015, and the elephants moved into a new, once again, state-of-the-art facility that not only quadrupled the space for the animals but organized that space in ways that simply was not conceivable even twenty years ago.

Anyone who investigates the history of keeping elephants in Western zoos cannot fail to notice story after story of inadequate facilities, inappropriate management techniques, psychologically damaged animals, endless stereotypic swaying resulting from boredom and stress, and the many, many deaths. Animals have been put down because they had become impossible to manage and because of chronic infections and musculoskeletal damage caused by unclean and hard flooring or insufficient movement. Others have been euthanized because of gastrointestinal problems and comorbidities associated with tuberculosis. Elephants have been stillborn or died young from physical injuries and a group of herpes viruses with an extremely high rate of mortality among new-born and young Asian elephants. Animals have suffered from decades of stress and poor management that has led to early deaths. But this is not the whole story. This is the partial story told by people seeking the removal of elephants from zoos. A more complete account is necessary. It would include accounts of keepers devoted to nurturing the animals in their care and of management practices that have radically changed as a result of new and frankly better ideas about how to meet the needs of elephants. It would include recognition of the professional, onsite veterinary care that zoos provide and the way they incorporate knowledge about how elephants live in the wild, exemplified, for example, in such innovations as facilitating births in the presence of other elephants, making it possible for elephants to spend time with the bodies of dead members of their community, devising systems that encourage animals to decide for themselves where and with whom they would like to be at different times of the day, and devising feeding systems that deliver food in smaller portions all around the exhibit so the elephants eat while walking instead of standing and binging on a bale of hay. It would include acknowledgment of zoo collection principles that have evolved from the idea that it is crucial to obtain one of every species and as many species as possible to the idea that it better to exhibit fewer species in more natural groupings and in larger exhibits, which for elephants means, among other things, trying to build multigenerational herds of female elephants. And it would include highlighting how zoos have adopted an approach that minimizes direct contact with the animals and assures physical barriers between keepers and animals, leading to a dramatic reduction in physical disciplining of the animals and an emphasis on positive reinforcement in training animals to be cleaned and stand calmly for veterinary and other examinations.

There were two young elephants at the zoo in the years we were visiting: Samudra, who was born on August 23, 2008, and Lily, who was born on November 30, 2012. Both were offspring of the male elephant Tusko, who had been on a breeding loan to the Oregon Zoo, and Rose-Tu, an elephant whose ancestry traced back to when elephants were first kept at the then Portland Zoo. Rose-Tu's mother was Me-Tu, and her grandmother was Rosy, who came to the zoo in 1953. Lily was born fifty years after Portland's—arguably North America's—most famous zoo elephant, Packy, who spent his almost fifty-five years of life in Portland and whose birth in 1962 was covered in an eleven-page article in *Life* magazine.[4] One could literally see parts of the history of elephant management on his body, throwing into relief how much better the world for most elephants in American zoos is than it used to be. The last time I saw Lily was in the summer of 2018. I was visiting my son who was living in Portland, and we went out to the zoo for the afternoon because I wanted to see her. Lily was out in one of the yards with her mother and another female, Shine, rolling about on a great mound of dirt and playing with water. She and the older elephants seemed utterly content, supporting what an elephant keeper at a zoo long ago told me, namely, that there is often nothing better for older elephants in a zoo than the presence of young elephants. Watching Lily with Rose and Shine and with her older brother Sam and another adult female, Chendra, demonstrated to me that making it possible for elephants to live in multigeneration families, as they would typically choose to do in the wild, can improve the quality of life for all the elephants in the group, even if, as critics of zoos argue, letting the animals reproduce is problematic because, to use Buffon's eighteenth-century language, it perpetuates a state of slavery.

Lily died on November 29, 2018, a day before her sixth birthday, from endotheliotropic herpesvirus. The onset was catastrophically quick, as is typical for the young Asian elephants who contract this disease. Despite every medical intervention, Lily died within a day of exhibiting initial symptoms of lethargy. There are undoubtedly people who think of Lily's short life and see only reasons for despair; they cannot imagine animals being in any way happy or content in a zoo. I think that the lives of animals in both "the wild" and in zoos are more varied and complicated than that. "The wild" is far from some kind of Eden, and so-called sanctuaries—so often seen as a solution for advocates who want elephants out of zoos—are clearly riven with their own problems when it comes to elephant care. The reality of such sanctuaries does not live up to the rhetoric about saving animals from the exploitation and captivity that spawned their creation.

It is true that the conditions of most elephants in major zoos in the West—not every elephant but a great many—are not as good as they could be, although the standards for keeping elephants have changed since Packy was born. It is now, for example, far from assumed that every major zoo should even have elephants in their collections. Unless the zoo is willing to commit to the extraordinary expense of building modern facilities and providing the extensive staffing required, the elephant program is ended. Fully closing down an elephant program can take decades because the zoo has to figure out what might be best for the animals currently in its care, but eventually, the program is shut down. This has happened in zoos around the world. While I don't think elephants naturally belong in zoos or are even particularly well suited to them, the fact is that, whether we like it or not, many elephants (some three hundred in North America alone) live in zoos, and so long as that is the case, we should try to continue to improve the conditions of their lives, just as we have over the many centuries we have kept members of the species in captivity.

The Good Elephant

On February 20, 1903, William Temple Hornaday, then director of the New York Zoological Park—more widely known both then and today as the Bronx Zoo—wrote to Carl Hagenbeck, the animal dealer in Hamburg, Germany, about acquiring elephants. The two men had been corresponding for several years, writing letters and sending telegrams to each other frequently, at times even weekly. Hornaday had come to rely on Hagenbeck as his principal supplier of animals for the rapidly growing zoo, which had opened in November 1899, and for Hagenbeck, not only was Hornaday an important client in his own right but also a connection that could potentially lead him to other American buyers. It was in both of their interests to nurture their relationship.

At the time he received this letter, Hagenbeck was fifty-eight years old and had been running his animal business for over forty years. Hagenbeck's father had been a fishmonger in Hamburg who had always found ways to make a little extra money. The story told for generations in the Hagenbeck family is that one day in the spring of 1848, when young Carl was just four years old, sturgeon fishermen who had been contracted to bring their entire catch also brought along some seals that had been caught up in the nets. His father put the seals in a basin of water in front of his shop and charged people to see the animals. The public interest was so high that he then sent the seals to Berlin; despite an uprising in the city, one of many across Europe during what became known as the Revolutions of 1848, Hagenbeck made a tidy profit from his exhibit. The exhibition of seals was followed by similar ventures with other animals that arrived at the growing port of the city, and in 1860, the then sixteen-year-old Carl took over his father's sideline interest, making it his own company, Carl Hagenbecks Thierhandlung, or "animal business."

By the 1880s, Hagenbeck's name was not only known in Hamburg but far beyond the borders of Germany, as he had become both the world's most successful exotic animal dealer and the organizer of extremely popular exhibitions of indigenous peoples, a traveling international circus, and much more. At the World's Columbian Exposition in Chicago in 1893, Hagenbeck erected a huge building on the midway to exhibit his animals and people, and by the time that he and Hornaday were corresponding a decade later, Hagenbeck was building his own revolutionary zoo in Stellingen, a suburb of Hamburg. The zoo, which opened in 1907, exhibited the animals in dramatic panoramas, and they were separated from each other and the public with moats instead of bars, a design that would become the model for exhibiting animals in zoos thereafter. The name Hagenbeck came to stand for a world of exotic ideas. As the German-born playwright Carl Zuckmayer put it in the late 1940s, "Hagenbeck is not a proper name, but rather, like Alaska or the Wild West, the expression of a mysterious, unexplored land, where one yearns for adventure."[5]

Ten years Hagenbeck's junior, Hornaday was forty-eight in February 1903 when he wrote to the German dealer about acquiring elephants. He had been founding director and curator at the New York Zoological Park since he was brought to the city in 1896 to lead the planning, building, and stocking of what would become, after only a decade, the largest and most spectacular zoo in the United States, an institution that easily rivaled any of the older and more established zoos of Europe or England.[6] For years before he came to New York and through to the late 1920s, Hornaday was a profoundly important American voice on issues of wildlife, conservation, and zoological gardens. His articles in popular magazines and newspapers reached an international audience, and his books, including *The Extermination of the American Bison*, published in 1889, and *Our Vanishing Wildlife* of 1913, stand today as milestones in the history of wildlife conservation. Hornaday picked out the proposed site of the new zoo in Bronx Park, and he developed the overall plan for the property, the design of the buildings, and the eventual collections. Beyond all that, he was responsible for the founding vision of the Zoological Park as an institution committed to science, education, recreation, and something altogether new, conservation. Because of his involvement with a wide variety of efforts beyond the zoo, though, he was also sought out from across the country for advice on all manner of issues relating to animals and their conservation. Letters to Hornaday could as easily come from an American president, a local bird-protection group in a small town in the South, or a guy in the Adirondacks who had managed to catch a bear and wanted to know what to do with it.

It is important to understand in considering Hornaday's role in the nascent animal conservation movement that he was not the modern stereotype of a tree hugger.[7] With his deeply nostalgic views about a better past, a past before all the "modern" ills he saw around him, Hornaday had much more in common with what we might now describe as a conservative politics. His efforts to stop the killing of birds for the millinery trade, save the last remaining bison, and regulate the hunt for fur seals were all about trying to stop what he considered the negative consequences of urbanization, industrialization, and immigration. In the end, Hornaday and his circle of associates were fundamentally focused on preserving their privileged positions in society, privileges they felt were part of the natural order of things. Like others figures in the New York Zoological Society such as Madison Grant, Hornaday deeply believed in what he would have called the "racial" superiority of northern Europeans and their descendants. He also believed that hunting large game was a way of preserving the vigor of the country. In the end, it was entirely consistent with his beliefs about his position in society—racially, professionally, and socially—that he would launch a highly public "war" against immigrants and people from lower working classes who left trash on the grounds of the zoo, that he would require visitors to secure his explicit permission before they took any photographs in the park, and that he would become obsessed with immigrants killing songbirds for food when much more damage to bird populations was being done by sport hunters.[8] It made perfect sense, too, that he would avoid the trend in zoological gardens in Europe of building animal houses reflecting architectural styles from around the world and would have all the zoo's buildings designed in the Beaux-Arts style typical of US municipal buildings of the period and that he would personally enjoy hunting and collecting trophy heads but would criticize others who lacked his scientific interest and cultural background for doing the same. Although animals were imperiled around the world, there was simply never a question of the value, in his view, of collecting living specimens of the rarest species for New York's (for America's) premier zoological collection.

Hornaday began his letter to Hagenbeck by ordering a collection of llamas, guanacos, vicunas, and alpacas, specifying that it should at least be as good as any collection anywhere else, that none of the animals be related to each other, and that they be exemplary specimens suitable to serve as gifts to the zoo from a "gentleman" who wished to make a donation. He provided two further instructions. He did not want any white llamas "as they always look so dirty," and he wanted the animals to be imported directly from South America so that there would be "no deterioration in their blood" because of inbreeding in zoos. From here, Hornaday turned to the main focus on his letter, purchases for a new antelope house, a major building he expected to open in the fall of 1903: "Our new Antelope House is going to require a large lot of expensive animals, and some of them will be of such rarity that they are not to be picked up every day. We will want an African elephant between six and seven feet high, and also an Indian elephant of a size to be determined." Apparently, a member of the society wished to give "something fine and very large" and would not consider giraffes or rhinoceroses.[9]

Over the following months, Hagenbeck and Hornaday returned to the issue of elephants. Hornaday repeatedly insisted that he was only interested in male elephants. On March 26, 1903, for example, he responded to Hagenbeck's news of an African elephant waiting for shipment in East Africa by saying that "if it is a male, reserve it for us; but if it is a female, you must look for a male for us. I know that tuskers are almost certain to become troublesome when they become quite old; but they are so much handsomer than females, I feel willing to take the additional trouble required in their management." In this same letter, he again urged Hagenbeck to also find a "fine, handsome, good-spirited, male" Asian elephant "between seven and eight feet high, for sale at a reasonable price."[10]

On April 9, Hagenbeck offered a female Asian elephant with a young calf, but Hornaday passed, reiterating that "at present, we must look out for showy males, with tusks that will grow very large."[11] On September 25, Hagenbeck offered the female and calf again, and Hornaday once again demurred: "The photographs show that the animals are undoubtedly fine and interesting; but for our purposes, we must have male elephants that will grow tusks,—even though they give us more trouble, and do not live as long as females." He continued: "I am hoping that your African elephant will soon arrive from Africa, and that in time you can get for us both species of African elephants, and I hope when you do that both will be males. Whenever you have the opportunity to procure a fine, good-tempered Indian elephant which is a tusker, and not too large to be introduced into a zoological park and trained, please let me know. I should say that the best size for us would be between five and six feet in height at the shoulders. Such an animal would be young enough that we could train it."[12] On February 23, 1904—a full year after the correspondence about elephants began and a couple of months after the antelope house had already opened, Hornaday again wrote to ask about a "tusker elephant seven feet high" because he must have one soon. Frustrated, he added that he could not "understand why all dealers should so persistently handle nothing but female elephants, when males are so much more interesting."[13]

Finally, on March 7, 1904, Hagenbeck sent the long-awaited news to Hornaday: "Your 7 feet tusker elephant is now on the way from Assam to Hamburg and if all goes well I hope to have him here about the middle of April, so that you can have him by about the middle of May in New York."[14] Hagenbeck also noted that he had a pregnant female coming as well and urged Hornaday to buy her for $2,500 and an additional $1,000 if the young elephant lived three days after it was born so that the zoo could have an entire "family." Hornaday bought the male and passed on the family. On March 24 Hagenbeck wrote to say that the "tusker elephant" would leave Calcutta on April 5: "My man writes he is a fine fellow, and I am pleased, I got the animal for you."[15] Then, on June 9, 1904, Hagenbeck sent a list of animals (fig. 4.2) to be shipped the next week from Hamburg to New York on the SS *Belgravia*, which included "1 male Elefant from Assam."[16]

On June 20, 1904, Hagenbeck sent news that the elephant had been shipped from Hamburg two days earlier; he also shared two new pieces of information

CARL HAGENBECK
Handels-Menagerie & Thierpark
Stellingen, Bez. Hamburg.

Telegr.-Adr.: Hagenpark Stellingen.
A B C Code 4th und 5th Edition.
Telephon No. 248, Amt 2.

Hoflieferant

Sr. Majestät des Kaisers und Königs.

Stellingen near Hamburg, June 9th, 1904.

W. T. H o r n a d a y Esq., Dr.,

Director of the Newyork Zoological Society,

Bronx Park,

N e w   Y o r k.

Dear Mr. Hornaday,

I got your telegram saying:

"Write immediately full list animals in next shipment for permit business."
I tried hard to get all your animals as well as mine for St. Louis away in s/s:"Patricia"leaving this saturday, but on account of the many passengers, this is impossible, and the Company promised me for sure, that the animals would leave next week in s/s:"Belgravia", in which I have the following animals for you to leave:

1 box: 1 male Elefant from Assam,
1 " : 1 Bakers Antelope,
1 " : 2 Aoudads,
1 " : 2 Snowleopards,
3 " : 5 Seals,
1 " : 1 female Rhesus with Baby, rare,
1 " : 1 Sheep /declared as an Antelope/

The following animals will be of mine in the shipment and if you think it necessary that we must also get a permit for these, I would be greatly obliged if you would also undertake to get the permission of entry for me:

Figure 4.2. Letter from Carl Hagenbeck to W. T. Hornaday, June 9, 1904.

about the elephant. First, Hagenbeck explained the arrangements for the keeper (referred to by the then-common term, "cornac" or "cornack") traveling with the animal and then paused to emphasize the particularly compliant behavior of the young elephant, using the word "docile" twice:

> The elephant-cornack, I send over to you, is a good fellow who knows well to handle him.—The elephant is very docile, & I am happy, that my man could exchange this one against that first one, that he bought, as this was such a brute, which killed his keeper in the presence of my traveller. But this elephant is as quiet, as very seldom male elephants are; the cornack had him out nearly every day in our garden, and it is as docile, as can be.—You can keep the cornack till the autumn; by that time I hope, you will have a man, who will take proper care of the elephant.—The salary of the cornack is $7.50 a month, during the time, he is on the journey, and during the time, he is with you in America, he gets $15.00 a month and his food. I think, he will sleep in the neighbourhood of his elephant.[17]

On July 5 Hornaday wrote that the animals had arrived safely, noting especially the elephant, "a fine, lusty beast, and while he has plenty of temper and disposition to mischief, I think we will be pleased with him." He noted that he did not "expect a male to be as quiet as a female" but concluded cautiously that "he is in the most perfect health, and within a month or so I am sure he will get settled down to his new life here and be quite contented. We shall also try to train him to behave himself." While Hornaday was pleased with the elephant, he was less so with the animal's attendant: "Just at present his Cornac is giving us about as much trouble as the elephant. He does not like the cooking at our restaurant, and we are about to erect for him, amongst the trees south of the Antelope House, a small house all for himself, in which he can cook to suit himself."

Then Hornaday changed tone, raising an alarming subject: "Yesterday, the Cornack in charge of the elephant told an educated Indian native that this elephant killed a man in India before he left for America! I would like to know positively whether this is true or not. If it is true, it is a serious matter. If it is not true, I wish you would give me the strongest documentary evidence that you can get from your traveller that the animal has never killed anyone." Hornaday, who was always worried about the reputation of the park and was probably specifically worried that his hopes to use the elephant as a riding animal might be dashed, explained that the man's statement might "make me some trouble with the public"; he recalled that Hagenbeck had mentioned that an earlier elephant had killed his keeper and hoped that the unwelcome news was simply the result of a mix-up. Before hearing back from Hagenbeck, Hornaday wrote again on July 11 with the news that the Indian keeper would be sent back to Hamburg on the ship *Pretoria*, leaving on July 16: "I have found it completely impossible to do anything with him. He is one of the laziest, surliest swine I ever saw come out of India, and is worse than useless. The trip to America has swelled his head so much that the quicker he gets back to his mud-hut at Johnpur [sic], the better for us all." Apparently, despite building the man a "small house" and outfitting it with cooking utensils and "food of every kind," the keeper refused to do sufficient work by Hornaday's estimation and instead demanded more compensation, insisting he would not do anything for less than $35 per month.[18] Hornaday suggested that Hagenbeck "get him to India in the least expensive way, and if you will put him in a swine-crate and send him, we will pay the expenses." "Our men can train the elephant perfectly well," he continued, and he noted that he had commissioned a saddle to be made after an example Hagenbeck had sent to Luna Park "so we shall soon be doing business."[19]

While this letter was making its way to Germany, Hagenbeck wrote back to Hornaday: "Do not pay any attention, what these black devils tell you. This elephant is, as I can assure you as a man of honour, as quiet as a cat." Insisting that the man had confused this elephant with the earlier one, he declared, "You cannot believe a single word of what these natives say. You must not make many circumstances with them. If you give them bread & butter, sugar & milk & perhaps a smoked fish, they are quite happy. If chickens are not too expensive, give your native one from time to time as well as some rice & curriestuff, and he'll soon become fatter. I must say, I never saw a fat Indian."[20] On August 17, Hornaday reported that "the elephant is getting on beautifully, and is now carrying visitors every day. His keeper gets along with him splendidly, and there is not the slightest trouble. The elephant weighs 3740 pounds, a very heavy weight for an elephant of his age."[21]

In a report to the members of the New York Zoological Society, Hornaday makes clear that after repeated efforts, including having to exchange one elephant who "at the last moment" "developed a bad temper" (he does not mention that the earlier elephant apparently killed someone), a suitable elephant had been obtained from Assam. According to Hornaday, Gunda, as the elephant was known, "has all the points of a 'high-caste' elephant," is a "good elephant," and "on Sunday, August 14th, began his regular work of carrying visitors. Keeper Gleason has trained him very successfully, and without the slightest trouble." He also makes his case for male elephants. "In captivity," he writes, "male elephants are much less common than females. In the hurly-burly of travelling-show life, male elephants are prone to resent the worriments that are common to all. The females are more patient and obedient under adverse circumstances, and therefore more desired. In the quiet life of a zoological garden or park, a male elephant has no excuse for being unruly; and as a zoological exhibit, a tusker is worth twice as much as a female. The price paid for 'Gunda' in New York, without his equipment, was $2,350."[22]

Gunda was put on exhibit in a specially designed stall with thick iron bars and reinforced walls in the new antelope house. It appears, however, that he pretty quickly started dismantling his stall and showing signs of violence.[23] One day, for example, he managed to knock down the partition to the next stall, which he then stormed into, startling Duke—the giant eland antelope that had been acquired from the Duke of Bedford. Early walks around the park were a failure, too, because he would wait for an opportunity and then bolt. But a story of redemption, one underscoring how he that had learned manners after he had come to New York, had clear news appeal. An October 1905 account in the *Washington Post* titled "Gunda, the Good Elephant" describes a mischievous elephant whose "heart is in the right place" and the young keeper, Frank Gleason, who was able to reach him through patience and devotion. "Gunda has the greatest faith and confidence in his young master," the article concludes, "and no terrace is too high for him to climb at Gleason's bidding. He has made rapid progress in the English language and understands many words. Besides being an unusually intelligent elephant, Gunda promises to become a monster of his kind."[24]

A half-page piece a few months later in the *New York Times* leading with "Dear Old Gunda the Elephant: He Wasn't Always the Amiable and Well-Behaved Pachyderm the Children Visitors to the Zoo in the Bronx Find Him Nowadays" echoed the *Post*. The article, which reproduces a large photograph of Gunda, his keeper, and children riding on the elephant's back, concludes: "It is a long, long time now since [Gunda] showed the slightest signs of ill-temper or moroseness, and he has developed into one of the most amiable and good-tempered elephants known in all captivity. What is more, he seems to enjoy [carrying children] as much as the children. It may be the exercise, which is so good and healthful for him, the change of air and scene, or it may be the peanuts, for which he puts up his trunk invitingly every time he has a new load on his back. He is the most friendly creature, and welcomes each visitor with a cordial invitation to shake hands, which, if somewhat clumsy, is kindly and well meant."[25]

The photograph in the *Times* was one of several taken apparently on the same day in the summer of 1905 of Gunda carrying children along the riding path in front of Lake Agassiz, near the main concourse entrance leading to Baird (now Astor) Court, the aesthetic focus of the park. The photographs were used for years in the zoo's *Popular Official Guide* written by Hornaday, were sold as postcards, and appeared in a variety of other publications. Postcard 1502D (fig. 4.3) is one of the set and shows three children on the back of the elephant, two facing the camera, each holding a doll and wearing summer dresses and straw hats, and another on the other side of the elephant with his back to the camera.[26] The photograph is exemplary of the kinds of photographs produced at major zoos at the time. A well-behaved elephant, an entirely appropriate animal to put children on the back of, stands calmly in a beautiful setting with his equally well-mannered and well-kempt keeper in a classic uniform. The photograph captures precisely what Hornaday seems to have had in mind when he began his correspondence with Hagenbeck about

Figure 4.3. Gunda in 1905. © Wildlife Conservation Society, reproduced by courtesy of the Wildlife Conservation Society Archives.

acquiring a riding elephant for the zoo. Hornaday had been specific in his order. He wanted male African elephants—one each of the East African savannah elephant and the West African forest elephant—and a male "tusker" Asian elephant that could be trained as a riding animal.[27] He was explicit in his use of the word "tusker" because many male Asian elephants do not grow tusks and tusks were critical for Hornaday: without them, his specimen would simply have been inadequate. It was for this latter animal that he had received support from a member of the New York Zoological Society, who turned out to be Oliver Hazard Payne, one of the wealthiest men of his day.[28] Hornaday needed something big (or at least something that would become big), something "showy," an elephant with tusks. He recognized that there was some risk in acquiring male elephants, but he felt the added bother would be worth it. For a time at least, he hoped for a calm, handsome elephant carrying children; he would also get a carefully contrived picture showing happy children, comfortable on the back of an elephant, and he would make certain that picture would get wide distribution.

The riding establishment at the park was set up at the beginning of the summer season of 1904 with the expected arrival of a suitable elephant in mind. The zoo acquired some small carriages and ponies, and two Bactrian camels were purchased in February 1905—the only disappointing feature of which was that they looked their best in the winter when no one was interested in camel rides. The main focus for the rides, however, became Gunda. As the park's 1904 *Annual Report* notes:

> The arrival, in July, of a fine half-grown tusk elephant from Assam, the gift of Col. Oliver H. Payne, greatly increased the interest of visitors in the riding animals, but the untrained condition of the animal, and the preparation of a suitable elephant saddle, occasioned some delay in the elephant service. The animal was placed in the hands of Keeper Frank Gleason, who from the very first has handled "Gunda" with excellent judgment and success. Within three weeks from his arrival in the Park, the elephant was carrying visitors, and unquestionably took quite a commendable degree of interest in his work. During his rather brief working season, Gunda carried 2,635 visitors, whose tickets yielded a total return of $395.25.[29]

In earning close to four hundred dollars in just over a month, Gunda accounted for more than half of the total receipts of $766.52 from the riding establishment for the entire summer.[30] The animal rides revenue in the next year almost doubled, bringing in $1,433.12. Revenues continued to increase the next year, and in its 1906 *Annual Report*, the park concludes that "the riding-animal establishment now has a well recognized popularity, and continually increasing patronage from children of all classes. Great care is taken to keep the ponies and vehicles up to a proper standard, and the attendants neatly uniformed."[31] That was the last season—and only the second full season—that Gunda gave rides to children.

The Finest Zoological Building in the World

The story of the mischievous elephant who reformed under the tutelage of a kind keeper began to fade as Gunda apparently became more and more difficult to manage. One of the first incidents happened in August 1906, when Gunda's "sweetheart," an elderly woman named Mrs. Lucretia Ann Hawes (fig. 4.4), who was a resident in the nearby Peabody Home for Aged Women

Figure 4.4. Mrs. Lucretia Hawes. Photograph by Elwin Sanborn, © Wildlife Conservation Society, reproduced by courtesy of the Wildlife Conservation Society Archives.

and who had been calling on Gunda regularly since he had arrived at the zoo, had a mishap with the elephant. On August 11, according to Keeper Gleason as reported in the papers, Gunda was "distracted by the heat" and sullen. When Mrs. Hawes arrived, though, he appeared to be pleased and, "extending his trunk through the bars, put the end of it against Mrs. Hawes' cheek. She petted the trunk. It was Gunda's way of kissing her," she said. "Then to the terror of those standing around, before Mrs. Hawes had any realization of her danger, the elephant wound his trunk around the old woman's neck, lifted her off her feet, and began to draw her over the brass railing toward the cage. In her struggle, one of Gunda's sharp tusks pierced a vein in the back of Mrs. Hawes's hand and blood poured from it." Apparently startled by the yells of the crowd and the keepers, "Gunda dropped his 'sweetheart' unconscious." "All afternoon," according to reports, "Gunda was inconsolable and would not let the children ride on his back as usual."[32] Mrs. Hawes survived the ordeal.

Then in July 1907, Gunda attacked one of the keepers. As reported, the events turned on a little box in Gunda's quarters in the antelope house, where the elephant placed coins. Gunda had been taught a popular trick: if a keeper or visitor threw a coin on the floor of his stall, he would pick it up, inspect it in his trunk, and deposit it in his box (called his bank) attached to a rafter above his head. He would then ring a bell and expect a treat from a keeper. The trick was a hit with the public, although many people apparently threw items other than coins in the hope that they could fool the elephant.[33] Soon it was discovered that Gunda had also begun to experiment with the stunt and was apparently removing coins surreptitiously from his bank. He would then pretend to find them as if someone had thrown them and ring his bell and expect a treat. When the keepers responded by putting nails in his box to stop him from retrieving the coins, Gunda started to pretend to drop the coins in his bank while actually storing them on top of a partition wall.[34] On July 28, a day like so many others, a child apparently threw a coin but it fell so far away from the elephant that Gunda could not reach it. According to reports, one of the keepers moved toward the coin with the intention of giving it to Gunda, but Gunda misinterpreted what the keeper was doing and seized him around the waist and dragged him into his enclosure. The keeper called out and two other keepers came running and prodded Gunda with their spiked goads until the elephant let go. Gunda had broken several of the keeper's ribs.[35]

The zoo minimized the event and reported that the especially hot day had simply put the elephant in a bad mood. An article in the paper on July 30 notes that "Director Hornaday is inclined to forgive [Gunda]. In the first place he realized that the heat of the last few weeks has affected Gunda more or less, and then he thinks Gunda merely yielded momentarily to an animal impulse and was probably sorry the next minute." Gunda, the article concludes, "seemed amicable again yesterday. His little eyes blinked kindly and he trumpeted in his usual joyful tone."[36] A couple of months later, however, Mrs. Hawes was again grabbed by Gunda and fainted, but the zoo tried to deflect attention away from the incident by putting out a story about the unusual friendship between the woman and the elephant.[37] On September 15, 1907, the *Times* published an article with the headline "Makes Pet of Big Indian Elephant: Aged Mrs. Hawes Feeds Him Cookies and Says She's Not a Bit Afraid." The article begins: "Probably many persons who visit the Bronx Zoo have halted in surprise before the elephant inclosure and marveled at the hardihood of a little white-haired woman who frequently stands close beneath the great swaying trunk of Gunda, a young Indian elephant, feeding him cookies from a big bag and talking to him and petting him fearlessly." According to the article, in the two years that Gunda had been at the zoo, the only real friend he had made was with the elderly Mrs. Hawes. Although the elephant had almost injured the woman several times with his "gigantic caresses," and the keepers were constantly afraid of something happening to her, apparently Mrs. Hawes had no fear of him and even when Gunda was "in a rage" at his keepers, he would "wind his trunk around her as he pleases."[38] A month later, under a four-column photograph of Mrs. Hawes and the elephant, the *Washington Post* followed the story that had been in the *Times* with "Captive Elephant's Love for a Frail Woman." The story related that although Gunda was first brought to New York to carry children," his "cross disposition long ago made that impossible." Apparently, the only person who could manage the elephant was Mrs. Hawes. Gunda and his friend," the article concludes, "are easily the star attraction at Bronx Park not only for visitors but for the members of the zoological society as well."[39]

During all of this, Hornaday purchased four more elephants for the collection. The zoo acquired three through Hagenbeck: a young West African forest elephant named Congo in July 1905 (fig. 4.5) and in June 1907, two very young elephants from the Sudan, Kartoom, a male, and Sultana, a female (fig. 4.6).[40] Finally, in September 1908 the zoo purchased a replacement ride elephant from Luna Park named Alice—the same Alice Helen Keller would meet fourteen years later. After her arrival, Alice quickly caused a stir by breaking away from her keepers and storming into the reptile house, smashing containers exhibiting various snakes and lizards along the way (fig. 4.7).[41]

Figure 4.5. Gunda and Congo. Photograph by Elwin Sanborn, © Wildlife Conservation Society, reproduced by courtesy of the Wildlife Conservation Society Archives.

Again, Hornaday downplayed the event to the press, emphasizing that Alice "was not vicious." "Not once did she try to do anybody any harm," he argued. "She was just scared out of her wits by her new quarters, her new keepers, and the absence of her three mates at Luna Park. She will be all right as soon as she gets used to her surroundings."[42]

Then, on November 20, 1908, the zoo opened its long-awaited and spectacular elephant house, a building "acknowledged to be the finest zoological building in the world," according to Hornaday.[43] The elephant house had been planned as the last major building of the garden, the completion of which would essentially mark the completion of the zoo after ten years of constant construction. At the opening, the building housed two Asian elephants (Gunda and Alice), two East African savannah elephants (Kartoom and Sultana), and one West African forest elephant (Congo), along with one Indian rhinoceros, two African black rhinoceroses, one hippopotamus, two South American tapirs and one Malayan tapir.[44] The white building clad in Indiana limestone

Figure 4.6. Kartoom and Sultana. Photograph by Elwin Sanborn, © Wildlife Conservation Society, reproduced by courtesy of the Wildlife Conservation Society Archives.

was centrally located at the end of the main promenade formed by Baird Court, and it divided the northern from the southern part of the zoo. It was 170 feet from side to side, 84 feet deep, and rose 75 feet to a dome and lantern covered in Guastavino tiles in a herring-bone pattern. In contrast to all the other buildings at the zoo, the elephant house was accessed at the center through tall arched doors on the long northern and southern facades (fig. 4.8). The early guidebooks make clear, though, that even though the park's "axial walk leads through the Elephant House," the building should not be used "*as a thoroughfare* for foot traffic," and park personnel had been "strictly forbidden to consider the walk through the building as a convenient highway."[45] On the north side of the building, the entrance was flanked by two stone African elephant heads and the head of a black rhinoceros, sculpted by the artist Charles R. Knight; the southern entrance was similar with the heads of two Asian elephants and an Indian rhinoceros created by Alexander Phimister Proctor.

The stalls for the animals were formed by single Guastavino arches and lit from above by skylights. Each stall was designed, according to Hornaday, "to display its living occupant as perfectly as a frame fits a picture."[46] At

Figure 4.7. Alice in the yard at the antelope house. Photograph by Elwin Sanborn, © Wildlife Conservation Society, reproduced by courtesy of the Wildlife Conservation Society Archives.

twenty-four by twenty-four feet square, the stalls were large compared to those in the antelope house, and each connected directly to a fenced outdoor yard for the animals. An article in the *Times* a couple of weeks before the opening declared, "The new dome-capped palatial elephant house in the Zoological Park, the most pretentious and fittingly constructed home for wild animals in existence, is finished, and will shortly open its doors to receive its colony of captives where they will enjoy more luxurious comforts, food, &c, in their roomy cages and outdoor quarters than they did in their palmy days in forest and jungle." Calling the elephant house "the most comfortably equipped building in the world for the housing and display of these large and intelligent creatures," Hornaday told the reporter almost wistfully, that "there is no pleasure in seeing a ponderous elephant chained to the floor of a small room, unable even to walk to and fro, and never permitted to roam at will in the open air and sunlight." In these outside yards, the animals "may

Figure 4.8. The elephant house, south façade. © Wildlife Conservation Society, reproduced by courtesy of the Wildlife Conservation Society Archives.

wander at will without the ability to harm any person or thing."[47] Hornaday expected that the building would be used "for at least two centuries," and in his communication to New York Zoological Society members in the *Bulletin* of September 1909, almost a year after the opening, he insisted that it is "clearly evident that the animals enjoy their cages."[48]

When the new elephant house opened in the fall of 1908, Hornaday felt that the zoo had accomplished a truly significant feat. A couple of small buildings—the zebra house and the eagles' aviary—remained to be built, but together they were, as Hornaday put it, "but a bagatelle, like the building of a garden summer-house for a stately mansion that is complete and occupied."[49] The elephant house was the finishing touch on a plan he had conceived a decade earlier, the fulfillment of a grand vision for what he was convinced would be seen as the most impressive zoo in the world.[50] As he puts it in the *Popular Official Guide*, "Of its buildings for animals, the Elephant House is the culminating feature of the Zoological Park, and it comes quite near to being the last of the series. In token of these facts, it is fittingly crowned with a dome."[51] Celebrating the three hundredth anniversary of Henry Hudson's discovery of what became known as the Hudson River, the hundredth anniversary of Robert Fulton's commercial steamboat service to Albany, and the tenth anniversary of the opening of the New York Zoological Park, the special Hudson-Fulton celebration issue of the *Zoological Society Bulletin* heralded the park: "The imperial City of New York presents to the world her Zoological Park, and invites mankind to behold in it a huge living assemblage of beasts, birds and reptiles, gathered from every region of the globe, kept together in comfortable captivity, and skillfully fed and tended, in order that millions of people may know and appreciate the marvels of the Animal Kingdom."[52] The elephant house, for Hornaday, was not only the architectural high point of the garden but also the most important building because of its remarkable and essentially unmatched collection of animals. With the animals in and the outside yards finally completed in the summer of 1909, Hornaday must have breathed a sigh of relief, knowing that he finally had the best possible facility for the animals—perhaps especially for Gunda. He would not have to worry any longer about the growing elephant's bad temper or his being a danger to the public or his keepers. As he concluded in the *Popular Official Guide*, "It is no wonder that dungeon-kept elephants go mad, and do mischief. If an elephant—or for that matter any animal—cannot be kept in *comfortable* captivity, then let it not be kept at all."[53]

Forever Chained to One Damned Spot

Despite the new building, Gunda's behavior did not improve. On July 29, 1909, just after the outside yards had been completed, Gunda attacked one of his keepers while they were both in the outside yard. According to the *Times*, the keeper, Walter Thuman was spreading straw when he heard Gunda behind him. He turned and saw that Gunda's "little eyes were angry" and that his trunk was raised. As Thuman ran, Gunda took a swing at the keeper with his trunk and charged. The keeper ran toward the building and then quickly turned and slid under a low rail at a spot in the corral where there was a little extra space. Gunda ran up, trampled the ground with his front legs and snapped six inches of a tusk off on one of the bars. Deciding, apparently, that the elephant needed to be disciplined, the keeper then got his whip and goad with its six-inch steel point and entered the corral facing Gunda. According to the newspaper report, "The elephant stood swaying his great body, swinging his trunk, watching him." The keeper advanced and Gunda turned away, retreating to a corner of the corral. "Thuman slashed him about the head with the whip, so that he tried to turn away, then he ordered him to lie down, and

Gunda lay down. Thuman, mounting his neck, rode him into the cage and dismounted and left Gunda quiet."[54]

This was not the end of troubles between Gunda and keeper Thuman. Three years later, on July 12, 1912, Gunda tried again. It was early on a Friday morning, before visitors were in the park. Thuman and another keeper, Dick Richards, were in the elephant house cleaning out the inside cages while the elephants walked about their outside yards. Richards was working in Alice's stall. Thuman, meanwhile, was in Gunda's when Gunda pushed open the iron doors to his outside corral. Thuman turned and walked toward the elephant with his goad to lead the elephant back outside. Then, "without a shadow of provocation" or a "trace of warning," Gunda swung his trunk at the keeper, hitting him with the "impact of a rushing locomotive" and sending him fifteen feet across the stall before the keeper hit the iron-clad wall. According to the *Times*, "the keeper was crumpled up on the cement floor, limp, dazed, and breathless," as Gunda prepared to rush at him. Somehow, the keeper managed to wedge himself tightly into the corner of the stall and Gunda could not reach him when he tried to crush him with his head. Gunda did, however, sink one of his tusks into Thuman's leg. Gunda tried to get to him again, but this time the tusk hit the cement, breaking it.[55] Hearing a commotion, Richards ran toward Gunda's stall, picking up a pitchfork along the way. He entered the cage and "lifting his pitchfork like a javelin, threw it with all his force at the elephant's head." The pitchfork glanced off the animal's head, and Gunda, "weakening, turned and shuffled out and into the corral." Thuman was then lifted to safety, while Gunda stormed around his outside corral "showing restlessness and anger at every approach." The afternoon papers ascribed Gunda's attack to excessive heat, repeating the reason given by the zoo three years earlier. According to the *Times* the next day, however, the park's veterinarian, W. Reid Blair, smiled, saying, "If there is one thing that an Indian elephant loves, it is excessive, Equatorial heat."[56]

This time, the decision was reached to chain Gunda in his inside stall from one back foot and one front foot, preventing him from turning on a keeper who might be cleaning up the animal's waste from behind. Almost two years later, on June 23, 1914, the *Times* took up the issue of Gunda's confinement in an article titled "Bronx Zoo Elephant Chained for 2 Years. Visitors Stirred to Pity for Gunda, Bound in His Cage after Attacking Keeper." Quoting from the sign before his stall that read "The average life of a wild elephant is probably 80 years," the article concluded that Gunda "can look forward to standing in the Bronx Zoo with one foreleg and one hindleg chained to the floor for the next sixty-two years." The article describes the relatively young elephant as looking more like "a battle-scarred veteran," with his broken tusk, ears "tattered and torn and pierced with holes where the hook has held him," and "his big skull" looking "worn, as though he had charged into stone walls to batter them down." The people "who flock to his cage and watch him as he tests the chains that bind him stand fascinated by the silent drama. He is the biggest elephant in captivity on this continent. He has more strength of bone and sinew than any other living creature. Yet he can move only two feet forward and two feet back, and then he drops his massive head and heaves from side to side, at once a grotesque and tragic figure."[57] Two days later Hornaday insisted that Gunda had simply come to the zoo bad. According to the *Times,* Hornaday believed that "some elephants" "seem to be born mean." When the keepers were interviewed, they suggested that Gunda's relentless swaying was simply his way of getting exercise.[58]

The paper's editorial staff responded to Hornaday's claim that Gunda's fate was the result of his "savage temper and a murderous disposition" by pointing to modern penology theories and insisting that "Gunda is in no degree responsible for his ferocity." The editors believed Gunda was suffering, insisting that his "life has been brought down to a mere existence." "His motions have been reduced to a monotonous swaying on feet practically fixed. There is reason for believing that his immobility has for him something of the deadly tedium that would soon drive mad a human being in like imprisonment." For the paper, the issue was only partly whether Gunda was suffering. An equally important issue was whether witnessing the animal's distress was an acceptable burden on the public. "The thought of him as he swings there ceaselessly back and forth," the paper declared, "is a horror by day and cause of bad dreams at night. Something must be done about it, and promptly."[59] The same day, the paper published three letters to the editor, all of which suggested that the management of the zoo was to blame for Gunda's behavior. The next day the paper published another editorial and five more letters from readers. One of the letters stated that "it is impossible for any human being with ordinary common sense not to see plainly the misery of this intelligent and powerful beast as it stands 'forever chained to one damned spot.' That this is an intolerable cruelty is unquestionably true." "If Gunda is incurably vicious, as it is asserted," the letter concluded, "then in the name of decency shoot him, and make of him a magnificent 'stuffed' specimen."[60]

On June 27, alongside yet another editorial and two more letters from readers, the *Times* printed a letter from Hornaday in which he responds defensively and apparently without restraint. "Ever since the Zoological Park was opened fifteen years ago," he explains, "I have warned the Zoological Society that the day would come wherein we would be attacked." Having expected criticisms to come from "natural enemies" like the SPCA, he shot back that he had never imagined that when the attack came it would be launched by one of the leading newspapers of the city. The paper had "started a movement that now seems very likely to lead to the shooting of Gunda at an early date, in response to the clamor of poets and editors and others who are so tender-hearted that they cannot bear to have him live any longer!" "Won't it be a spectacle for gods and men," he scathingly asks, "when Gunda, munching hay in the most comfortable animal room in the whole Zoological Park, hears the roar of the elephant rifle that sends a steel bullet crashing through his skull and brain in the name of The Merciful, The Compassionate?" Hornaday states that he feels those attacking "his management" should be forced to watch Gunda killed; as for himself, he insists that he will "ask to be excused." After Gunda was finally shot, though, he adds, he planned to erect a plaque that would read "Shot to death in response to the demands of tender-hearted friends who wanted him to be free to turn completely around in his room." With his close to forty years of studying the "minds of elephants" and his fifteen years of leading zoological gardens, Hornaday believed he was the only person sufficiently expert to decide what would be best in this situation. He called those seeking to set Gunda free from his chains "mental incompetents" and insisted that there were only two options for Gunda: leave him alone and hope that his period of fury would pass or shoot him where he stood.[61]

On June 28, the paper published another editorial with a sketch of Gunda chained in his exhibit, along with another three letters proposing such ideas as building a new exhibit and sending him back to "Africa"; another editorial and letter appeared on June 29. Letters and editorials continued to appear into July, with letters being published almost every day. On July 19, just days short of a month since the troubles with the paper had begun, the *Times* devoted an entire spread to letters from readers under the widely distributed photograph of Gunda carrying children from 1905 (fig. 4.3) and announced breaking news in the story. The headline read "TIMES READERS PROTEST AGAINST GUNDA'S IMPRISONMENT: Young and Old Write Letters Urging That the Chains Be Taken Off the Big Elephant at the Zoo and Plans Are Being Made to Give Him Some Freedom." "Gunda will not have to 'serve time' in his cage much longer," the article explains. "The readers of *The Times* have won a pardon for him." According to the paper, the zoo had decided to build a sliding door that could be operated from outside Gunda's cage that would "permit Gunda to pass from his cage to his yard without coming in contact with his keepers. When this door is completed the chains will be removed from Gunda's legs, and for the first time in two years he will have an opportunity to stretch his 9,000 pounds of tired bone and muscle." The article claimed that when Hornaday issued his ultimatum that the choice for Gunda was simple—he must be either chained or shot—the newspaper's readers simply refused to accept either of the alternatives. Pointing to letters by the hundred, the editors insisted that Hornaday and his supporters were ignorant of other, more modern ways of thinking about punishment and captivity. Twenty-four letters were published in the spread. The letters were "uniformly courteous," according the editors, in addressing the director of the park, and they also provided substantive contributions, including the idea of the sliding door. One letter quoted Hornaday's proclamation in the *Popular Official Guide* that "there is no pleasure in seeing a ponderous elephant chained to the floor of a small room, unable even to walk to and fro, and never permitted to roam at will in the open air and sunlight."[62]

The public had spoken and, apparently, Hornaday was finally persuaded to give a remarkably simple solution—a sliding door—a try. Over the next weeks, though, occasional writers to the paper lamented that little seemed to be changing at the park. By the middle of August, the park had returned to its old line: Gunda would be released from his chains when he became safe. According to a statement that was issued under the auspices of the New York Zoological Society but that very much had the tone used by Hornaday, while many of the letters that had been sent to the paper and the society made clear the writers' "well-meaning desire to relieve Gunda from what they believe an unnecessarily close confinement," many were "quite hysterical" and demonstrated only that the writers were simply not aware of the facts of the case. "They all clearly exemplify," the statement continues, "the present tendency of the man in the street to undertake the solution of a problem which requires a very considerable experience with animals in general and elephants in particular, as well as an exact knowledge of the underlying facts."[63] In other words, Hornaday knew best and the public was simply inadequately informed and therefore not in a good position to make suggestions about how to improve Gunda's circumstances. The paper's editorial column lit up again the next day, expressing surprise that, despite assurances, Gunda's situation had not improved and that, moreover, no improvement was in sight. Accusing the New York Zoological Society of going back on earlier commitments, the paper raised the stakes, vowing that "the campaign, if necessary, can be begun again, and, should its resumption be compelled, the temptation will be strong to enter more deeply than before into the whole question as to justification for keeping any wild animals in captivity."[64]

A trickle of letters continued to show up in the paper, but then on September 19, the zoo announced that Gunda's period of violent behavior had passed and that a new solution had been devised—a fifty-foot-long cable had been stretched across the ground of his outdoor enclosure and Gunda attached to the cable with a metal ring, so he could move from his indoor cage to his outside yard (fig. 4.9). Apparently, though, the elephant continued to just sway in his interior stall unless he was prodded to go outside. A letter to the editor published on October 3, almost two months after the beginning of World War I, laments that "in the bitter strife between the nations of the world Gunda seems to have been almost forgotten. Even though he has been allowed fifty feet additional chain, will not the humane people of our city make another earnest appeal . . . to release him from those cruel chains and the humiliation

Figure 4.9. Gunda and the cable. Photograph by Elwin Sanborn, © Wildlife Conservation Society, reproduced by courtesy of the Wildlife Conservation Society Archives.

and animal torture to which he is subjected?" A letter published on December 27 claimed that essentially nothing had changed for Gunda and asked plaintively whether anything else could be done. On January 12, the *Times* reported that Gunda was back to having one foreleg and one hind leg chained to the floor. "All day long," the paper states, "the huge animal stands swaying, moving his great body in a diagonal direction—weaving the zoologists call it." Hornaday reminded reporters that Gunda had been given more liberty but that he preferred to stay in his stall and sway. A statement issued by the society points out that the executive committee had determined that "so far as can be determined [Gunda] does not resent his captivity provided he is left alone to sway his great bulk from side to side." Claiming that if the elephant were unhappy, he would not eat, the statement notes that "Gunda eats voraciously as ever: from which it may be fairly inferred that he is reasonably contented."[65] An editorial the next day wondered out loud about what happened to the promise to free Gunda, but the story's time had clearly passed: the public's attention had turned to the war in Europe.

Six months later, on June 22, 1915, Gunda's life came to an end. He had been at the zoo for eleven years. Along with a picture of Gunda standing with Keeper Thuman, the *Times* on June 23—one year to the day after the beginning of the paper's efforts on behalf of the elephant—published an article titled "Bullet Ends Gunda, Bronx Zoo Elephant. Dr. Hornaday Ordered Execution Because Gunda Reverted to Murderous Traits." The lead reads: "The troubles of Gunda, the bad elephant of the Zoological Park in the Bronx, ended yesterday morning, when from the vantage point of the small iron door leading from the side of the elephant house into his inclosure, Carl E. Akeley, Assistant Curator of the American Museum of Natural History, sent a bullet crashing into Gunda's brain."[66] The article reports that Gunda had refused food for several days, that "for the first time during the weary months it has been necessary to keep him again chained up, and that that Hornaday had finally become convinced that Gunda was suffering. Hornaday concluded that Gunda "was no longer any use as a zoological attraction." "In the last weeks," the paper notes, "he was a demon and it was no longer safe for any one to go into his inclosure." At 8 a.m. on that Tuesday morning, after the other elephants had been moved to their outside exhibits and before the zoo opened to visitors, Akeley stepped into the elephant's stall with an elephant gun. Gunda stood, chained.

> He glared at Mr. Akeley for a second, and then, straining at his chains, he made a vicious swing with his trunk at his executioner. Mr. Akeley is a distinguished

> elephant hunter, and as Gunda lifted his trunk he raised his rifle and fired. The steel bullet caught the elephant between the eye and the ear, penetrating his brain. The uplifted trunk stopped in the middle of a sweep, the animal closed his wicked little eyes, opened them and then, quivering, sank down and died practically without a struggle. Director Hornaday and Keeper Walter Thuman went out while the animal was being killed.

If the news account has an almost slow-motion quality, Hornaday was clear, despite having not been present, that the "shot produced instantaneous paralysis of the brain. The huge beast simply dropped in his chains and died without a struggle." Backtracking what he had been saying for a year, Hornaday explained that Gunda "was suffering from the restraints of captivity and his desire to kill men increased. So, because the animal was not enjoying life and did not seem likely to do so throughout more than six months of each year, it was a kindness to kill him." There was apparently still value in the animal, though. According to the article, in addition to the ivory of the tusks and the meat obtained for the lion house, "Mr. Akeley figured that he had on hand 250 square feet of the finest kind of elephant skin, worth at the present wholesale price about $9 a square foot."[67] His skin alone, apparently, was "worth" $2,250—a hundred dollars less than his purchase price.

A Grotesque and Tragic Figure

Hornaday had understood from the very beginning that in acquiring a male elephant, he was perhaps setting the park up for difficulties in the future. The problem can at least partly be traced to the basic physiology and way of being of male elephants. Even today most elephants in zoos and those remaining in circuses are females. Until very recent decades, in fact, it was an absolute rarity to find a male elephant in a zoo; the few that were kept in zoos, though, tended to get noticed. First of all, elephants are highly sexually dimorphic—male elephants can grow to be twice the size of females. People who have only seen female elephants can be stunned when they see a full-grown male elephant. When Gunda was shot, at about seventeen years old, he likely still had some growing to do. At nine thousand pounds he was big, but I have met male Asian elephants in zoos exceeding fourteen thousand pounds, and male African elephants can grow even larger. The size and power of male elephants is related to the way they live and to how they become successful sires of further generations. Elephant herds are matriarchal, consisting usually of multiple generations of related females. As male elephants mature in their teens, they begin to leave their herds and to start living more and more alone or with "bachelor herds" for periods. When a male is fully mature and able to compete with other males, a significant part of his life is spent keeping track of dispersed female herds, looking for breeding opportunities with females in estrus. While the female herd is organized around social hierarchies and personal connections and maintains close affiliations with a wider social circle of other herds, male elephants live in social worlds that include familiarity with different herds of females and relationships and competition with other males. While females may be content to stay in areas where there is plentiful food and water and where they feel safe, males are essentially designed to traverse long distances and overcome almost any obstacle in their quest for breeding opportunities.

Adult male elephants experience periods of highly aggressive and unpredictable behavior that tend to be annual and seasonal but can be more frequent. The periods are called "musth" and are related to although possibly not driven by hormonal changes (including massive increases in the production of testosterone). Musth varies in its length for different elephants but can last for months at a time. During musth, adult males can and typically do become violent, and most captive elephants experiencing musth (in zoos, work camps, and other settings) are isolated and maintained in reinforced quarters. Hornaday, who was familiar with descriptions of musth, was willing to take on the risk of acquiring male elephants because, as I have noted, he felt that they were simply more spectacular than female elephants. When Gunda started to become violent, the zoo was quick to conclude that this was only a seasonal problem for the animal and that it would pass. As curator Raymond Ditmars put it in the *Zoological Society Bulletin* in 1915:

> [Gunda] is of high caste, a patrician among elephants, and is wilful and desperate only at specific times. There is a period each year when most adult male elephants are more or less disturbed. This occurs in the spring, and the breeding period is designated as "musth." The maturing Gunda indicated this condition in the spring of 1913. He had been daily becoming more surly and one morning when Keeper Thuman was leading him out of the stall where the animal had been at liberty, Gunda charged, hurled Thuman to the floor and badly gored him with one of his tusks.[68]

Hornaday had used the term "high caste" to describe Gunda from the moment the animal was purchased. It was an expression that seems to have been used frequently and with little clarity in the West during this time by various people who claimed to know a great deal about Asian elephants. In his 1867 study *The Wild Elephant and the Method of Capturing and Taming It in Ceylon,* Tennent tried to provide a more complete treatment of the expression. Pointing to a Singhalese work on the natural management of elephants, Tennent lists the marks of "inferior breeding": "eyes restless like those of a crow, the hair of the head of mixed shades; the face wrinkled; the tongue curved and black; the nails short and green; the ears small; the neck thin, the skin freckled; the tail without a tuft, and the fore-quarter lean and low." In contrast, "high caste" was marked by "softness of skin, the red colour of the mouth and tongue, the forehead expanded and hollow, the ears broad and rectangular, the trunk broad at the root and blotched with pink in front; the eyes bright and kindly, the cheeks large, the neck full, the back level, the chest square, the fore legs short and convex in front, the hind quarter plump, and five nails on each foot, all smooth, polished, and round." High-caste elephants, Tennent notes, are immensely rare, but owning one would "impart glory and magnificence to the king."[69] By this description of ideal conformation—probably the best-known account available to Hornaday at the time—Gunda was almost anything but "high caste." For Hornaday, though, it seems to have been critical that the elephant be officially described as "high caste"; that way, of course, Gunda could "impart glory and magnificence to the king" or at least to the City of New York and its zoological park and perhaps also to the donor, Oliver Payne.

Gunda was not a patrician among elephants. He does seem to have been experiencing musth, however. As Hornaday puts it in his 1915 *Annual Report,* "The Indian elephant developed quite early in the year his annual fit of 'musth,' and while we had hoped that it might be less severe than usual, it proved to be more so. His bad temper was so pronounced and dangerous, and his rage at his keepers so constant, it became evident that at last old Gunda was suffering from the confinement that was necessary to keep him even measurably under control."[70] Musth, however, cannot wholly account for Gunda's apparently uncontrollable behavior.

In zoos and circuses over most of the twentieth century, the normal practice was to have the keeper or trainer insert himself into a social hierarchy in which the elephants were the dominant figures. The keeper or trainer then used the goad—I use the word "goad" because it was common at the time, but it has had many names over the centuries, including "ankus" and, more recently, "guide"—to make himself the dominant figure instead. Elephants in herds in nature establish their rank within the herd through their physical presence, using their bodies, sometimes aggressively, but more generally through a broad array of gestural, sound, and scent cues. The keepers at the New York Zoological Park, who were familiar with breaking and training horses, relied on their prior experience with other large animals and were convinced that the way to manage elephants was, essentially, to convince the elephant that they were more powerful and more dangerous than the elephant. Keepers who were especially adept at maintaining their position did not have to strike the elephants—or at least only had to do so rarely—and so the hope of the keepers was that the animals would essentially concede the battle early on and accept their status. Many elephants did, and this is why when Gunda first assaulted Thuman in July 1909, the keeper responded by immediately reasserting his control over Gunda by forcing him, with his whip and goad, to lie down and then carry the keeper back into the house.

But most male elephants, including Gunda, did not entirely concede to keepers in the struggle for dominance. Gunda's hard wiring thus in part explains why he was a management problem for the zoo. It makes sense that there would have been an almost constant struggle between the keepers and Gunda for dominance because Gunda demonstrated a particular hatred for his keepers, evident in how violently agitated he would become when his keepers were heading his way. This is the lesson, I think, of one of the last official photographs taken of Gunda. The elephant is lying down, Thuman sitting on him with his goad across his legs as an audience looks on (fig. 4.10). To the public, the image may have seemed to reflect a calm and relaxed friendship, but in fact the picture captures a single moment in a long-standing conflict, a moment when the human keeper had the upper hand.

Beyond musth and the basic nature of male elephants, other factors certainly contributed to what the *Times* called "the progress of [Gunda's] madness."[71] His capture as a young elephant, his travel to Calcutta, his shipment to Germany and then New York, his years in the antelope house, followed by more years of living in limited quarters with relatively brief interactions with a small circle of social relations (the keepers and figures like Mrs. Hawes), and the staggering paucity of intellectual and physical engagements for him must have had a profound impact on Gunda's basic psychological makeup. There is little doubt that the keepers and Hornaday endeavored to meet Gunda's needs, but they had a devastatingly incomplete understanding of what an intelligent, physical, and active entity like the growing Gunda needed. Hornaday's noting that Gunda seemed to enjoy giving rides to children may simply have been an attempt to promote the park, but it also possible that Gunda did in fact enjoy and look forward to the social engagement and peanut treats of being a ride

Figure 4.10. Gunda and Thuman, 1915. Photograph by Elwin Sanborn, © Wildlife Conservation Society, reproduced by courtesy of the Wildlife Conservation Society Archives.

animal. But with the end of his activity as a ride elephant, even that "enrichment" ended. That Gunda clearly spent most of his young adult days swaying or weaving, displaying what now would be called stereotypic behaviors, suggests that his life experiences and environment likely had a deep impact on his mental health.

But there is more. In a letter to the editor of the *Times* on June 24, 1915—two days after Gunda was shot by Akeley—a reader named M. E. Buhler recalls that "the last time I saw Gunda, Saturday, some twenty or thirty school children, laughing and noisy, were gathered about his cage. His annoyance was extreme. Thrusting his trunk out at them once or twice, he turned abruptly away, standing for a moment with his face to the wall. He then commenced the old pitiful nosing at the iron bars of the gate that led into the yard. He is outside now, and peace be with him!"[72] The letter points to the role of Gunda's audience in his life and difficulties. The accounts of Gunda make it clear time and again how fascinated the public was with both the elephant and his violent behavior. In its opening barrage on the park over the treatment of Gunda, the *Times* notes how people would "flock to his cage and watch him" as he tested "the chains" that bound him. He was, the *Times* perceptively observes, "a grotesque and tragic figure."[73] Of course, many—and maybe even most—of those visitors were not content to witness Gunda's trials in silence. Instead, the audience would do what it could to aggravate and enrage Gunda because then the exhibit would be even more spectacular. The classic moves here would be to pretend to throw food or to throw nonedible objects into Gunda's mouth. Gunda learned early on to raise his trunk when visitors came to see him; Mrs. Hawes was not the only visitor to the gardens who carried a big bag of food to throw to the animals. But as Gunda matured and began to exhibit violent behaviors, his spectacle changed as well, and no doubt many visitors enjoyed provoking him into fits of aggression.

Elwin Sanborn, the official park photographer and the photographer responsible for most of the images reproduced in this chapter, wrote a piece for the *Zoological Society Bulletin* in September 1912 in the wake of the attack on Thuman that resulted in his hospitalization. He observes that "the visitor can make even a friendly animal dangerous." "Gunda has been for years a center of interest. Because he can throw back his head at the beck and call of every man, woman and child while they heave all kinds of food into his eager throat, and chase up and down the fence in rage when he is tormented, he has become a great attraction." Even those whose hearts are "overflowing with the milk of human kindness" and who simply feed the animals because they feel the animals need the food can cause them problems. Gunda, he wrote, "is like the majority of men and women. He has moods. He has good qualities, and his bad ones are not improved, either by ceaseless baiting or misdirected attention from people who imagine that he never gets a meal."[74] Although the park promoted the story of Mrs. Hawes and the elephant, she was clearly, for Sanborn, part of the problem. Gunda likely eagerly anticipated her visits but was probably also frustrated when she left and he found himself once again the object of taunting or passive neglect by the public. This is not meant as an indictment of Mrs. Hawes or those at the zoo who encouraged the relationship; the benign story simply shows the opportunities and inevitable limits of the New York Zoological Park in the early twentieth century.

Dying in a Zoo

Dying at a zoo is not at all an unusual occurrence for the animals who live there. We do not hear about it in the news, but almost every animal at the zoo will die there, at another zoo, or in transit between zoos. There have been high profile cases of animals that have been reintroduced into their natural ranges—Arabian oryx who today wander the deserts of the Middle East, condors who fly in the American Southwest, and Przewalski's horses who roam in Mongolia, all descendants of former zoo animals and all living now in variously managed versions of "the wild."[75] Over the last couple of centuries, too, many animals have escaped from zoos, and some of those succeeded in living for extended periods before being recaptured or dying from one cause or another. An American raccoon escaped from a Hagenbeck transport in the late nineteenth century, and people claimed to see it running about the Lüneburg Heath in northern Germany for years longer than raccoons normally live; a Chilean flamingo, who came to be known as Pink Floyd and who had flown away from a municipal aviary in Salt Lake City, split his time for seventeen years migrating between the Great Salt Lake and Montana.[76] But these are all exceptions. Just months, then, after Carl Akeley came to the zoo to shoot Gunda, he showed up once again carrying his elephant gun. The still young elephant Congo, the second elephant to come to the New York Zoological Park, had apparently been suffering an inflammation of nerves in his forelegs leaving him in severe pain and unable to walk. On Wednesday, November 3, 1915, Akeley shot Congo, and his remains were sent to the American Museum and to the College of Physicians and Surgeons.[77] Congo's death was not reported in the newspapers.

Gunda's death was obviously tragic. In the end, he was killed because of inadequate facilities, insufficient resources, and incommensurate empathy and imagination. His fury and rage were the result of his nature and the conditions of his life. His behavior, though, was exacerbated by the public that came to see him and taunt him. His heightened rage then led inevitably to further measures to control him, which led to more rage, the public becoming both more fascinated and outspokenly sensitized to his captivity, and ultimately to his execution. All of this contributed to the dramatic tragedy of Gunda's death, but the tragedy was not just his death but also his life. Telling Gunda's story is not about condemning zoos or condemning Hornaday, Gleason and Thuman, Hawes, the editorialists of the *Times,* Akeley, Hagenbeck, or of anyone else. I am telling Gunda's story because it is part of the reality of the modern engagement with elephants. Not every day of Gunda's life was a living torture. He had good days. Even in his last years, he undoubtedly had days of calm and quiet, days of positive interactions with other elephants and people, days that were comforting and filled with pleasant smells, tastes, and sounds.

Certainly, too, other elephants' lives at the Bronx Zoo, other elephants' lives at the hundreds of zoos around the world over the last century, were not as difficult as Gunda's.[78] While it is sadly obvious that facilities for elephants in zoos were for many years woefully inadequate for the flourishing of these remarkable animals, it is also obvious that the newer, larger, more complex and interesting elephant exhibits that have been built in recent decades are significant improvements for the animals, for the staff, and for the public.

I have had the opportunity over the years to be at zoos after hours, after the commotion of the day has died down. I have also visited zoos on days that look like the opposite of the perfect day, a rainy or bitterly cold Monday morning when there are fewer visitors. Off-exhibit spaces can also be remarkably peaceful and calm areas. At a zoo near where I live there is an off-exhibit area for African hoof stock where the animals live during nights and long winters. The stalls there are peaceful, the light is muted, and the air is sweet with fresh hay. In the off-exhibit areas in the Oregon Zoo's old elephant house, I spent quite a bit of time just listening to the building. Elephants can be very quiet animals, something nineteenth-century hunters often note in their memoirs. In the old Oregon building, there were muted sounds of a radio playing classic rock, the swishing of a trunk gathering up sawdust to throw over a back, and the chips of house sparrows looking about for nesting materials. Occasionally I would hear a low rumble from Packy or one of the other elephants, or a squeak, or, more rarely, a trumpet or roar. I would catch casual conversations between keepers talking about moving an elephant to a different yard or just chatting about the day. All these sounds were comforting. But other senses came into play as well. Ever since the introduction of large public zoos in the nineteenth century, visitors have complained about the smell of certain buildings—especially the cat houses. When I have visited behind the scenes in elephant houses over the years, I have often heard from other guests that the smell is comforting. Elephant houses have traditionally been called "barns," and the domestic feel of these places, even the newest ones, recalls a shared living space with animals.

I am sympathetic to some of the criticisms of zoos, but I also find them to be so important to the worlds of both humans and animals that dismissing them as irrelevant or as anachronisms seems to me to simply miss the point. It is true that there are plenty of crying children and plenty of people who seem to barely notice anything between stops at various food stands, but there are also people who spend time trying to learn all they can about the animals and their lives, who do not tire watching the prairie dogs, and who visit specific animals every time they come to the zoo. And the animals often recognize those people, too.

The people who work directly with the animals in zoos do so because they feel called to. This is not work that is highly paid, and it is often barely noticed by people visiting the zoos. This is not a nine-to-five job; these people are passionate about the work they do and passionate about physically caring for the animals. This means cleaning up the mess, preparing food, interacting with the public and answering the same questions over and over again with a smile. In the case of elephants, the work is intellectually, emotionally, and physically demanding, and although these keepers are often the targets of criticism, they show up every day and do their best to maintain the animals' health in circumstances that are often challenging. I admire them, and I am grateful for the work they do. And, of course, the reach of zoos themselves often goes beyond concern for the animals directly in their care. In this respect, it is important to recognize that year after year, the Bronx Zoo's parent organization, the Wildlife Conservation Society, provides an immense amount of support to field research and conservation, including more than half of the around $200 million dollars contributed by all American zoos to wildlife conservation efforts around the world annually. As long as elephants continue to live in zoos, I think we should support the work of those who have committed themselves to caring for the animals and also support the institutions that work to improve the circumstances both of the animals in their direct care and those living in the wild.

A Descendant of Mastodons

In the early summer of 1913, a story circulated in local newspapers in Kansas, Louisiana, and Mississippi about an unusual event that had taken place in an arena in Mexico.[1] The version of the article that appeared in the *Lyons Republican* of Lyons, Kansas, on June 3, 1913, led with the headline "Bull in Fight with an Elephant: Queer Combat Is Described by an American" and is filled with the sort of misinformation one gets used to with a topic like this. The article quotes a Mr. H. F. Lang of Philadelphia, who claims to have been walking down Mesa Street in El Paso toward San Jacinto Plaza when he heard "the familiar strains" of the Sousa tune *Invincible Eagle* being played by an approaching band. At the end of the procession, he recalled, there was a large elephant draped in a canvas painted with the message "This African elephant will fight a ferocious bull from Chicucha to the death in the bull ring in Juarez tomorrow, Sunday, February 10. Price for admission, $200, box seats; $150, shade seats; $100, sun seats." February 10 was actually a Monday, "Chicucha" was presumably Chihuahua, and those would have been some very expensive seats! An advertisement preserved in the Municipal Archives in Seattle, Washington, corrects the article. The fight between the bulls and an Asian elephant named "Ned" took place on Sunday, February 2, 1913, with box seats going for $1.50, shade seats for $1.25, and sun seats for $1.00 (children were half price).[2]

According to the newspaper account, the fight began when Ned was led into the arena and chained to a stake by one of his hind legs to prevent him from reaching the public. A bugle sounded and a bull entered the ring through a gate, stabbed at the last moment with a banderillo, or barbed dart, a method of enraging bulls before a fight. The bull "ran around the ring once or twice and finally saw the elephant, and stood stock still, sizing up Mr. Elephant. The elephant also saw Mr. Bull at this time and they both stood staring at one another." At this point, the article continues, the bull ran away, and the crowd called for another bull. This bull, too, ran and jumped a five-foot fence. The bull was rounded up and prodded back into the ring, this time stuck with a "rocket banderillo," a lit firework that began to sputter on his back. The bull saw the red ribbons tied to the elephants' neck and tail and charged. Ned parried by squatting down and when the bull struck him, the force "knocked the bull over." Over the next hour, the article relates, the bull continued to charge and attempt to jump out of the arena. A bullfighter then entered the ring with his cape and killed the bull with a sword to its heart. It had, in the end, been a battle to the death, although the audience most likely had hoped to see the bull killed by the elephant rather than a man.[3]

This is just one account of this event. Most of the versions I have seen were written some twenty years later when the elephant in the story died and articles began to appear across the United States relating "remarkable" episodes in his life. Details vary: dates and the number of bulls change, and even whether any of the bulls actually charged the elephant is disputed. Lee Clark, the "son" of M. L. Clark & Son's Circus, claimed five bulls were sent into the ring against Ned and that none of them would fight. The audience was incensed, the police confiscated the film that was taken of the event, and the circus was fined $500 because no fight took place. Clark related, though, that he avoided jail because someone had to take care of the elephant, and in the middle of the night he slipped back across the border with Ned and just never went back to pay the fine.[4] We cannot, in the end, know what truly happened that day; all we have are different accounts told for different reasons. Lang's account, though, is the sort of story one could tell without having seen anything more than an ad for the planned event. Indeed, "H. F. Lang" may simply have been an invention of a writer who had heard a story, wanted to report it in the paper, and came up with a "source" to tell it.

Ultimately, the details of the event do not matter much anyway. It was not the first time, after all, that an unusual confrontation had been staged between animals; these sorts of events have drawn crowds for millennia.[5] Even today, people who play the computer game *Zoo Tycoon* will drop a lion in with the zebras or introduce an elephant into the midst of a crowd of people to see what happens, and the programmers have developed their algorithms in anticipation of these impulses: lions just start killing zebras and people start screaming and running away from elephants. Whatever happened that day in Juárez, though, it is at least clear that the event was organized and staged. But why, and why did people go, and what made the story a "story" for the newspapers? People may have attended because they were curious about how an elephant and a bull might respond to each other in an arena; others may have been drawn by the aesthetics associated with the "art" of bullfighting. I think most of the audience simply went to see an elephant and also to see whether it would be frightened by a bull or not.

An illustration of the event (fig. 5.1) that was often included in newspaper stories reporting Lang's account depicts a rather diminutive, Horton-esque

Figure 5.1. "Made a Deliberate Charge," *Syracuse (KS) Journal*, June 6, 1913.

elephant appearing more surprised than anything else at the snorting charge of a bull, bleeding from a banderillo in its shoulder. Hardly the image of the "monster elephant" promised in promotions, the illustration has a humorous tone that derives in part from the ribbons the elephant is wearing on his tail. And this points to another aspect of these kinds of spectacles. Laughter was part of what was anticipated. An illustrator could just as easily have portrayed a massive, resolute elephant watching a bull trying desperately to exit the arena by jumping over a fence, but the moment that people wanted to see—and that the illustrator, it seems, hoped to capture—was the moment of direct contact, which manages to epitomize the absurdity of the whole enterprise. The fight was an obvious mismatch, and the cartoonist presented it as, essentially, a contest between a surprised, champion heavyweight facing a physically outclassed upstart who did not know he had no chance of winning. In 1942, amid the success of Disney's *Fantasia* (1940) with Ponchielli's "Dance of the Hours" and *Dumbo* (1941), the Ringling Brothers and Barnum & Bailey Circus presented an elephant ballet with music by Igor Stravinsky, choreography by George Balanchine, and fifty elephants in pink tutus along with fifty ballerinas. At one level, like the bullfight, the elephant ballet with all the training, props, and hype was clearly taken seriously, but the result of the performance was also undoubtedly comical. The idea of elephants dancing, just like the idea of an elephant squaring off with a charging bull, was intended to bring a smile as much as it was designed to draw the public's attention. At one level, the goal of both events was a comic fail.

A Mud Show

By 1913, Ned had already been with the Clark Circus for more than a decade, but information about those early years is murky. It appears that he was imported to the United States from what is now Thailand by the New York animal trading company of Louis Ruhe in either 1901 or 1902, although an inquiry in 1932 resulted in a letter from the company saying it did not have any record of such an elephant but that the lack of a record did not mean it had not imported him.[6] It is also commonly claimed that Ned was already twelve years old at the time of his import, although the circus historian Homer Walton, relying on sources that included Lee Clark, reported in 1958 that Ned was about five feet tall and only about five or six years old when the Clark Circus acquired him, which would have meant that Ned was only about three years old when he was originally imported. This timeline seems plausible because it was always easier to ship a very small elephant than a large one. Ned was initially purchased by a William F. Smith, proprietor of a show known in 1901 as the Great Syndicate Shows and in 1902 as the Great Eastern Shows. After the 1903 season of what Smith then called Howe's Great London Circus, he decided to sell his business, and M. L. Clark went to Kansas City where Smith had property and purchased the elephant and some horses.[7]

The roots of the M. L. Clark Circus reach back to the mid-1880s when Mack Loren Clark, who was born in 1857, put together a wagon show called the Clark Bros. Shows with his older brother Wiley.[8] The venture closed in 1891 when the brothers failed to make payments on equipment, but M. L. seems to have started again with a small medicine show that traveled from town to town presenting minstrel comedies and selling elixirs for whatever complaints people might have. By 1895, the show had apparently grown enough that M. L. decided to purchase a Bactrian camel and a small Asian elephant from Hagenbeck. The story goes that the animals were shipped from Hamburg and were picked up by M. L. in Mena, Arkansas, and so he named the elephant Mena.[9] M. L. had set up his home and the winter quarters for his circus in Alexandria, Louisiana, as the circus toured mostly through the Southeast US, beginning the season in late February or March and running as late into the fall as possible. In 1903 the show had a single ring, but a second ring and larger capacity tent was added in 1904. By 1907 the show had been rebranded as the M. L. Clark and Son's Combined Shows and Trained Animal Exhibition, his son Lee having joined the operation.

Typically, the show would pack up and leave immediately after the last performance in the evening and arrive at the next town either during the night or early in the morning. If a full night of travel was needed, the wagons might stop along the way; the horses, camels, and elephants would be tethered or hobbled, and everyone would have a chance to catch a little sleep. Advance men would have already visited the town to secure permits, rooms, and provisions, and advertisements would have been posted to drum up interest in attending the circus. When the wagons finally showed up the excitement would build. Men and boys from the town would be hired on the spot, the pay for which for the boys was a ticket to the show, to help ready the grounds, raise the tents, and feed and water the animals, although the elephants were essentially always taken to the water, except when the circus wanted to amuse local children or when a kid holding a pale of water out for an elephant might provide a good publicity shot. Soon, it would be time for a parade or at least a procession of a few animals (especially the elephants) through the town, and then in the afternoon, the circus would open. Even circuses traveling in wagons could become quite large. By 1910, the M. L. Clark and Son's Circus moved with more than sixty wagons, eighteen cages for animals, over two hundred horses, and eight camels. There was a 120-foot round-top main tent, large tents to house the performing and draft horses and mules, and an advance team traveling on a half a dozen buggies and wagons.

The elephants—Ned and Mena—would walk along with the wagons and the camels from town to town.[10] The M. L. Clark and Son's Circus is remembered as one of the biggest and last of the wagon shows, often called "mud shows" because they traveled on dirt roads. Although the circus did pick up an occasional truck and tried rail for at least a couple of seasons, it always came back to the wagons, the dirt roads, the horses, and the small towns. If trains made it possible for the larger circuses to travel greater distances to visit just the big cities where they could have multiday stands, the method of travel of M. L. Clark and Son's Circus meant it could be in a new small town every day. The circus was always on the move, and there was a lot of walking and work to be done by humans and animals alike. Around the circus grounds, Mena and Ned worked as tractors raising the tent poles, pulling up the canvas, and moving wagons around by pushing with their heads or pulling in harnesses.

There is a photograph of Ned and Mena from sometime between about 1915 and 1921 on the road with the Clark show (fig. 5.2). In the background of the photo can be seen some hitched horses, a wagon, and a Bactrian camel.[11] At this point, Ned still had some growing to do, but he is clearly already a massive elephant with powerful shoulders and long, slender tusks. The photographer shows the two elephants standing with a circus worker beside his Paint stock horse in front of a hitched wagon. The photograph has an almost spur-of-the-moment feel, but it is clearly a carefully composed shot—it is meant to pique curiosity about the traveling life of a circus and the extraordinarily unusual animals and people walking around the countryside of the American South. The animals are shown relaxed and calm, and the photo gives the impression, with the positions of the elephants, the young man, and the horse, that it has been collaged together. Whether it was or not, the intention was to show what the circus looked like when it was packed up and ready to move between towns, and the photo does that fairly well. Ned, with his left front leg forward as if he is testing his chains, is shown anchored closely to Mena, and both elephants have thrown dirt and straw across their backs. With all the wagons, horses, caged animals, camels, elephants, and performers getting ready to move out of town, I can only imagine that there was a great deal of sound, dust, yelling, and effort to manage hundreds of not-always-steady

Figure 5.2. Ned and Mena. Photograph by F. E. Halek, William "Buckles" Woodcock Jr. Collection, http://bucklesw.blogspot.com.

horses. This photo, though, has a quiet quality, as if the animals are content and relaxed.

In the early years with the Clark Circus, Ned's act had quite a few different features. He would walk on the tops of driven pegs or bottles, stand on a tub and turn around, stand on his hind feet or front feet, lie down and then sit up. He could do a little waltz-like dance and, according to Walton, he was one of the early elephants to do a headstand.[12] Mena's act was not that different. She could walk around the ring on her knees and stand on her two right feet or two left feet. She could stand on her hind feet and sit on a tub. With these basic "tricks," the trainer could then put together a little show in which he would narrate a story while the elephants walked around the ring between the various parts of the story. Mena performed with the M. L. Clark Circus until the outfit was sold to the E. E. Coleman Shows in the fall of 1930, a few years after M. L.'s death. In the late 1920s and '30s she was often advertised as "the largest elephant in captivity." In the 1930s she continued to perform in various truck shows connected in some way to the Coleman operation including Duggan Bros. Circus in 1934, the Bailey Bros. Circus in 1935, Johnny J. Jones

Shows in 1935 and 1936, and the Jack Hoxie Circus in 1937.[13] In 1940, Coleman sold her to the Al G. Kelly & Miller Bros. Circus, in whose care she died on October 25, 1943, at the end of the season, at about fifty-five years old. Mena undoubtedly had many unusual experiences during her life. On one notable occasion, for example, a truck driver navigating hilly country in Pennsylvania for the Duggan Bros. show in July 1934 lost control of his vehicle; he managed to jump out of his cab just before the whole truck, with Mena chained inside, crashed down a forty-foot embankment.[14] Just the number of miles she must have walked across her more than thirty years with the Clark Circus is astounding to think about, not to mention all the people who saw her when they were kids and then saw her again when they were adults bringing their own kids to the circus. A story on the front page of the *Journal-Gazette* of West Plains, Missouri, on May 31, 1928, clearly exaggerated when it claimed that "Mena, the big elephant, has walked more than 250,000 miles," but she certainly did travel a very long way.

As Ned got older he became more difficult to handle. Over time he performed less and less, and his tricks fell by the wayside, but he was still useful as a drawing card because he was quickly becoming one of the largest elephants in North America. When moving from town to town, Ned would be chained to Mena to prevent him from wandering off, and in the lots, he was chained to driven pegs and sometimes hobbled with chains, especially when he was in his periods of musth. A reasonable argument has been made by the retired elephant handler and circus historian William "Buckles" Woodcock Jr. that all the walking and other work kept elephants like Ned out of mischief. As Buckles noted once, "As a rule, those males that had to walk overland were more interested in finding a nice place to lie down and sleep during the day rather than bother anybody. Once they were transported by rail and received an abundance of hay and grain it was another story."[15] Still, as the years went by, it appears that hauling along an increasingly unpredictable large male elephant through the back roads of Texas, Louisiana, Arkansas, Kansas, and elsewhere became a significant concern.

The Mightiest of Living Creatures

Perhaps it should not be a surprise then, that in the middle of the 1921 season, Ned was sold for $6,000 to the Al G. Barnes Circus, then headquartered in unincorporated land the circus owner called Barnes City near Culver City, California. On July 3, 1921, according to Richard J. Reynolds, Lee Clark had Ned crawl into a railroad baggage car because he was too big to stand in one, and the elephant was shipped from Seligman, Missouri, to the Barnes Circus, which at the time was traveling in Minnesota.[16] Barnes then had a special train car built for the huge elephant with the floor dropped down between the trucks so that he could stand while being shipped around the country by rail. Whether it was because he no longer walked between towns or because he had just grown bored and frustrated with life in the circus or simply because he was becoming a mature male elephant with all the hormonal and other aspects that process entailed, over the 1920s Ned evidently became an increasingly dangerous and unpredictable presence with the circus. As Reynolds puts it, "He became a roguish, rough-house bull, and his rampages were legendary."[17]

Though some of those who worked with him undoubtedly continued to call him Ned, Barnes also changed the elephant's name to Tusko, a name he must have thought more suited to a creature with six-foot tusks. The Barnes Circus soon began marketing its new elephant as the "the mightiest of living creatures," "the last living link between civilization and the glacial ages," and "the last of the giant race of prehistoric mammoths" (fig. 5.3). Barnes claimed that Tusko came in at more than thirteen feet high, weighed twenty thousand pounds, and was the largest elephant ever held in captivity. In coming up with those numbers, it seems he wanted more than anything else to surpass the old declarations by P. T. Barnum that his famous elephant, Jumbo, who died in 1885, was the largest elephant ever exhibited. Barnum claimed that Jumbo was twelve feet high and weighed fourteen thousand pounds, so Barnes made certain that Tusko was taller and heavier—at least in print.[18]

The first of Tusko's legendary rampages came on May 15, 1922, just ten months after Barnes had bought him, when the circus was visiting a small town along the Skagit River in Washington State, about sixty miles north of Seattle, called Sedro-Woolley. As in the case of the Juárez bullfight, there are many different versions of this story. In recollections typed up in 1931, Barnes describes Tusko during his early days with his circus as a "big, good-natured clown" who loved the spotlight and getting dressed up. According to Barnes, "Elaborate trappings were made for him of ermine, plush, satin, and silk. His blanket was a gorgeous affair. Old Tusko always rumbled happily when 'dressed up,' seeming to enjoy the magnificent spectacle that he made." Noting the "fancy howdah, fit for an East Indian maharaja" that would be placed on Tusko to carry the circus owner, Barnes recalled that "At the opening of the performance I appeared on Tusko, introduced by the announcer in loud, sonorous, impressive tones as 'Al G. Barnes in person.'" Tusko, Barnes felt,

Figure 5.3. 1922 Al G. Barnes Circus herald, a type circus poster. Circus World Museum, Baraboo, Wisconsin.

"seemed to know that he was the center of attraction, and he took slow, measured steps, making his ponderous stride one of dignity."[19]

There is a publicity photograph of Barnes sitting in his howdah on Tusko. The elephant is anchored with chains to another elephant, Ruth, while the trainer, named "Mississippi" Nance, uses the blunt end of bullhook to signal to Tusko to raise his trunk (fig. 5.4). Chains run from the elephant's tusks through loops at his feet to a chain going over his back and connecting at his chest. Other chains restrict the movement of his legs. Barnes, sitting grandly atop an elephant who was clearly not going on any rampage during this photoshoot, wants to be seen as relaxed and casual, as if sitting on top of a ten-foot high elephant was just a quotidian detail in his grand life. In a post on his circus blog describing another photograph of Tusko and Barnes, Buckles Woodcock pens a line that could caption this photograph well: "Mr. Barnes positioned as tho he knew something."[20]

Figure 5.4. Tusko and Ruth with "Mississippi" Nance and Al G. Barnes about 1923. Circus World Museum, Baraboo, Wisconsin.

In his 1931 recollections, Barnes argues that what caused Tusko to get in trouble in Sedro-Woolley was simply that Barnes himself had to go to California for business and Tusko missed him and went looking for him. According to Barnes, the public expected to see him riding on Tusko, so when he was called to California, a substitute named William K. Peck was instructed to get on Tusko and impersonate Barnes. "At the beginning of the fourth performance," according to Barnes, with Tusko kneeling with all his "trappings" and a ladder against his side, Peck started climbing the ladder. When he was about halfway up, "Tusko suddenly rebelled. He jumped to his feet, throwing Mr. Peck and the ladder aside, and began backing up, swinging his head and trunk from side to side." Tusko bellowed with rage and "people scattered in all directions." Tusko went out the exit, threw cars out of his way in the parking lot, and headed down the street toward a dance pavilion. Tusko then swerved and crashed through a pen sending "fluttering and squawking chickens" through the air. With trainers, horses, other elephants, and a large number of locals in pursuit, Tusko then pushed a two-story house off its foundation, crashed through a garage of another house, investigated a pigpen, and then headed for the hills in a "trumpeting rage, covered with foam and lather, destroying all in his path." He stormed through an apple orchard, leaving it looking like a "cyclone had struck it," and then disappeared in a pine forest. There, according to Barnes, Tusko discovered a still and started in on the "whisky mash." A fixer for Barnes followed the trail, paying off residents for their losses and getting their signatures on releases. "Stimulated by the alcohol and reveling in his freedom," according to Barnes, Tusko enjoyed the "beautiful warm night" and "kicked his heals in the air like a gamboling lamb of gigantic proportions, stood on his head, and went through various contortions. It was a comical sight to see the intoxicated big fellow, still dressed in his howdah and fancy trappings, as he played like a kitten, appearing in the dim light as some fantastic dream of a dressed-up doll. The earth vibrated for several hundred feet around him as he performed his antics." The next morning Tusko was retrieved, and Barnes recalls that when he finally got back to him, "old Tusko threw his trunk about my body and drew me close, rumbling and talking as if to assure me that he had been willing to wreck the world to find me again. I talked to him and fed him raisins, peanuts, popcorn, and candy, and he embraced me with a loving tenderness as he munched his sweets."[21]

Like much of what Barnes was reported to have said in newspapers and elsewhere, his recollections of the "rampage," related almost a decade after the events and less than a year before he died, seem designed more than anything else to create a tall tale that also emphasized his own importance in the world. An article that appeared two days after the events in papers across the country, including in the *New York Times* with the title "Elephant on Rampage: He Leaves Thirty-Mile Trail of Destruction in Washington State," was more succinct:

> TACOMA, Wash., May 17.—Tusko, described as the largest elephant in captivity, is reported in a special dispatch to The Ledger today as peacefully consuming his fodder with a circus at Bellingham, Wash., after an afternoon, night, and morning of rampage that stretched for 30 miles from Sedro Woolley, Wash.
>
> Tusko hurled his keeper, H. Hendrickson, 30 feet in the air, breaking several of his ribs, and then proceeded through the streets of Sedro Woolley, capsizing three automobiles and turning a dance into a riot. Then he headed for the hills.
>
> Flattened fences and orchards and calls from excited farmers and loggers betrayed Tusko's line of flight to several hundred men and boys in pursuit. At one logging camp Tusko uprooted three telephone poles.
>
> A farmer, looking out of an upper story window, gazed upon the elephant's mighty back hunched in an unsuccessful effort to overturn the house. A barn proved less stanch, and after breaking in Tusko ate his fill and then proceeded onward.
>
> At dark Monday the several hundred pursuers made camp in the woods, taking up the trail at daybreak yesterday. It was in a valley known as The Garden of Eden that Tusko apparently returned to normalcy, as calmly and as suddenly as the spirt of rampage had possessed him. Sauntering up to two other elephants that had been included among his pursuers, Tusko meekly permitted his recapture.[22]

It was probably more like three miles than thirty, and there was no one working for the circus named "Hendrickson," but this story seems to have the basic elements of the tales about the rampage that became more embellished over the decades, as is common with such events and the stories told about them.[23] The Sunday magazine section of the *Salt Lake Telegram* for June 25, 1922, for example, included a full-page article that also appeared in other regional papers with the page-wide headline "What Happened When the Elephant 'Took a Notion.'" A large and reasonably accurate illustration of Tusko depicted with a dozen people fleeing before him and a trail of destruction behind drew readers into a story that was short on facts as to what happened that day and even shorter on facts about the history of this particular elephant. Noting that "there is something suggestive of mysterious power in the captive elephant," the article describes an animal whose "age is reckoned well along in the hundreds" captured four years earlier in India and secured at the cost of $100,000—a figure, the story claims, arrived at by simply halving the amount the circus proprietor claimed to have paid—an animal that was known as the "king of the jungle and noted from Bombay to Lucknow as the wildest man-killer who ever defied big game hunters."

By the time of the *Telegram* article, the costs of the damage caused by Tusko in Sedro-Woolley, which had likely amounted to a few thousand dollars at most, had been inflated to $75,000.[24] Of course, this sort of publicity was a boon for gate receipts. Barnes, in fact, worked to keep his circus in the newspapers, placing ads and, it seems, providing ready-made articles to local papers.[25] It is clear, too, that Barnes recognized the publicity potential of a dangerous elephant. A brief notice on June 26, 1922, with the headline "Elephant Star of Parade: Tusko a Drawing Card for Al G. Barnes' Circus that Shows Monday Evening" in the *Lincoln Journal Star* makes plain enough that a mad elephant could help a circus. According to the article, "Many spectators at the Al G. Barnes circus parade followed the procession down O street several blocks Monday noon, attracted principally it appeared, by Tusko, the monster elephant. Tusko's immense size and reputation as a bad actor fascinated the crowds. The circus was billed for two performances at the grounds on South street."[26]

Sedro-Woolley was not the only town the elephant rampaged that summer. The *Evening News* on August 9, 1922, in Harrisburg, Pennsylvania, put an all-caps headline *above* the newspaper's front-page banner that read "ELEPHANT ATTACKS KEEPER." According to the story, circus staff, referring again to the ill-fated "Harry Hendrickson" who was once-again hurled to the ground, breaking more ribs, lost control of Tusko while unloading him at the train station. Before the staff could regain control an hour later, the elephant had smashed headfirst into a sleeper wagon, literally ripped up some rail to which he had been anchored, and then ran toward a "panic-stricken crowd" before knocking down a fence and eating all of Dr. O. A. Newman's beets at his house on 617 Race Street.[27] In the end, Barnes wanted Tusko to be traveling with the circus, leading the parades and in the spectacle, or "spec," at the beginning of the shows, even as he realized that the elephant was becoming increasingly difficult to handle. Barnes's early publicity photographs emphasized the size and tusks of Tusko. A series taken in 1923 at the new Al G. Barnes Circus Winter Home and Zoo in Barnes City, for example, showed Tusko standing on tubs, saluting with his trunk up, and Barnes standing on the elephant's back, balancing himself with his hand on a building with cages for menagerie animals (fig. 5.5). The tubs and the ladder were then cut out and the cages replaced with a stretch of three-story buildings along a main street; the resulting pictures were used in publicity to magnify Tusko's already extraordinary size (fig. 5.6); they magnified (literally and figuratively) Barnes, too, and that was also part of the point of these sorts of images. Tusko was clearly one among a very long line of elephants, including Jumbo in the 1880s and even the elephants that belonged to Henry III in the thirteenth century or Charlemagne in the ninth, who were meant to display the power of their owners and were kept simply because they impressed.

Indeed, the way Tusko was presented and exhibited in the early twentieth century echoes plainly enough the way the very first modern elephant exhibited in North America was shown at the end of the eighteenth century. On April 13, 1796, Jacob Crowninshield of Salem, Massachusetts, one of five brothers captaining ships sailing back and forth to India, Africa, and Europe, docked his ship *America* in Manhattan and unloaded an elephant.[28] Crowninshield had acquired the young elephant, "almost as large as a very large Ox" for $450 with the hope of selling it for $5,000 when the ship arrived in New York.[29] Despite a voyage of four months, the elephant arrived apparently healthy, was sold, and was successfully exhibited for a number of years, walking from town to town along the American East Coast.[30] A broadsheet advertising the appearance of the elephant in Boston in the fall of 1797 is headed with an illustration of a notably strange-looking beast. Although an effort was made to get the toenails about right, the back legs seem pulled from a hippo and are missing an elephant's characteristic knees, the long tail makes little sense, the ears and head are out of proportion, the body looks a bit like that of a giant pig, the trunk suggests an earthworm, and the tusks look like something one might find on a wild boar (fig. 5.7).

The illustration seems implausible, but the advertisement sought to establish the credibility of the exhibit in the opening sentence with a reference to Buffon's work from just decades earlier. The text reads, "THE Elephant, ACCORDING to the account of the celebrated BUFFON, is the most respectable Animal in the world." The text highlights the animal's "knowledge" with an anecdote about the elephant remembering its keeper after ten weeks of absence. The broadside describes the elephant as being four years old and

Figure 5.5. Tusko standing on tubs, and Barnes standing on Tusko, 1923. Circus World Museum, Baraboo, Wisconsin.

weighing three thousand pounds and notes that it would not reach its "full growth till he shall be between 30 and 40 years old." The animal evidently consumed 130 pounds of food every day and "drinks all kinds of spirituous liquors; some days he has drank 30 bottles of porter, drawing the corks with his trunk." The elephant was reportedly tame and had never attempted to hurt anyone; he had also been seen at the "New Theatre" in Philadelphia "to the great satisfaction of a respectable audience." Making clear what sorts of

Figure 5.6. Barnes and the mighty Tusko, advertisement, 1923. Circus World Museum, Baraboo, Wisconsin.

audiences would be welcome, the advertisement alerts readers that "a respectable and convenient place is fitted up at Mr. Valentine's, head of the Market, for the reception of those ladies and gentlemen who may be pleased to view the greatest natural curiosity ever presented to the curious, and is to be seen from sun-rise, 'till sun-down, every Day in the Week, Sundays excepted." Adults were to be charged a quarter dollar and children nine pence;

THE

Elephant,

ACCORDING to the account of the celebrated BUFFON, is the moſt reſpectable Animal in the world. In ſize he ſurpaſſes all other terreſtrial creatures; and by his intelligence, he makes as near an approach to man, as matter can approach ſpirit. A ſufficient proof that there is not too much ſaid of the knowledge of this animal is, that the Proprietor having been abſent for ten weeks, the moment he arrived at the door of his apartment, and ſpoke to the keeper, the animal's knowledge was beyond any doubt confirmed by the cries he uttered forth, till his Friend came within reach of his trunk, with which he careſſed him, to the aſtoniſhment of all thoſe who ſaw him. This moſt curious and ſurpriſing animal is juſt arrived in this town, from Philadelphia, where he will ſtay but a few weeks.——————He is only four years old, and weighs about 3000 weight, but will not have come to his full growth till he ſhall be between 30 and 40 years old. He meaſures from the end of his trunk to the tip of his tail 15 feet 8 inches, round the body 10 feet 6 inches, round his head 7 feet 2 inches, round his leg, above the knee, 3 feet 3 inches, round his ankle 2 feet 2 inches. He eats 130 weight a day, and drinks all kinds of ſpiritous liquors; ſome days he has drank 30 bottles of porter, drawing the corks with his trunk. He is ſo tame that he travels looſe, and has never attempted to hurt any one. He appeared on the ſtage, at the New Theatre in Philadelphia, to the great ſatisfaction of a reſpectable audience.

A reſpectable and convenient place is fitted up at Mr. VALENTINE's, head of the Market, for the reception of thoſe ladies and gentlemen who may be pleaſed to view the greateſt natural curioſity ever preſented to the curious, and is to be ſeen from ſun-riſe, 'till ſun-down, every Day in the Week, Sundays excepted.

☞ The Elephant having deſtroyed many papers of conſequence, it is recommended to viſitors not to come near him with ſuch papers.

☞ Admittance, ONE QUARTER OF A DOLLAR.——Children, NINE PENCE.

Boſton, Auguſt 18th, 1797.

BOSTON: Printed by D. BOWEN, at the COLUMBIAN MUSEUM Preſs, head of the Mall.

Figure 5.7. Description of the Crowninshield elephant, 1797, in Frank Cousins, broadside advertisement of the first elephant exhibited in America, ca. 1865–1914. Frank Cousins Collection of Glass Plate Negatives, reproduced by courtesy of Phillips Library, Peabody Essex Museum, Salem, MA.

all visitors were warned not to come near the elephant with "papers of consequence" because the animal appeared to have a habit of grabbing and destroying such items.[31]

Diarist Elisabeth Sandwith Drinker reports that on hearing that this elephant was in Philadelphia on November 12, 1796, she "immediately concluded to see it." However, it turned out that the elephant was not going to be shown at the Chestnut Street Theatre, the "New Theatre" referenced in the Boston advertisement, after all. That venue appears to have been mentioned in the Boston ad only as part of a more general effort to suggest the respectability of the show. Instead, Drinker had to follow an alley near a market to "a small and ordinry room, where was tag. rag &c [sic]." "The innocent, good natured ugly Beast was there," she writes, and "it is indeed a curiosity to most that sees it, one of the kind never having been in this part of the World before." According to Drinker, the elephant did little more than stand in the room and drink the alcohol it was given. "I could not help pitying the poor Creature," she states, "whom they keep in constant agitation, and often give it rum or brandy to drink—I think they will finish it 'eer long."[32] Despite Drinker's report, this elephant likely did have a few tricks beyond the apparently always crowd-pleasing uncorking of a bottle. At the very least, she probably rang a hand bell, picked up and handed to her keeper coins thrown to her, and held up her trunk in salutes to get treats thrown into her mouth.

Of course, many more complex tricks have been taught to elephants over the hundreds of years since the Crowninshield elephant. Young elephants, especially, but older elephants, too, learned tricks like balancing on a large ball held between rails, rearing up and walking on their back feet, doing handstands, swirling around on top of a tub, putting their front feet up on the back of another elephant and walking in what came to be called a "long mount." Elephants were trained to make pyramids with two elephants lying down and a third elephant standing on the backs of the bottom two, front feet on one elephant and back feet on the other. Elephants were taught to walk the "tight rope"—two rails set into tubs a few feet above the ground that the elephants could walk down using just their back feet or their front feet. Special tricycles were made for elephants that they could peddle around performance rings. Eventually, elephants were trained to do the truly remarkable "one-foot stand" in which the elephant would first do a handstand on a tub and then lift one of his or her front legs and balance on just one leg.[33] And, of course, much could be done dramatically with comical and at times astounding costuming, with a trunk, and with a good trainer. Elephants performing "barbershop" numbers, playing music, or blowing water at clowns became classic numbers. The audience was supposed to be both amazed and amused by all of these acts.

As impressive as Tusko was even just walking down the street (fig. 5.8), the elephant's days of traveling with the circus were numbered. Barnes explains that "his love for me remained apparently unchanged, but the old sunny disposition had departed." Tusko "grew unruly," Barnes recalls, "indulging in

Figure 5.8. Tusko leads the Barnes elephants, ca. 1923–25. Circus World Museum, Baraboo, Wisconsin.

more stampedes, and attacking his trainer at the slightest opportunity. There were many narrow escapes, and Tusko began to be considered a 'bad' elephant."[34] Increasingly, and especially during periods of musth, Tusko would be taken off the show and sent back to the Barnes Circus Zoo in California, where he was put in a specially designed pen and visitors could pay twenty-five cents to see him.[35] According to Barnes, "Huge channel irons were used as posts, embedded in concrete, with railroad irons as fence rails, spanning about 8 feet from post to post." Concocting another background story for Tusko, again entirely disconnected from the truth, Barnes suggests that Tusko's strength and his proclivity to violence owed partly to his heritage: "He was employed at handling logs in the lumber camps of Tibet when I first heard of him through an animal dealer, who sent me his measurements; these showed him to be the largest elephant ever captured, and he was reported to be the largest ever seen in the world. He stands higher and weighs over 2 tons more than the celebrated Jumbo. From his general characteristics, I am satisfied that he is no ordinary elephant, but that he breeds back to the mastodon strain." Not surprisingly, when this "mastodon" was released in the pen, "he immediately began to test the strength of the posts and railroad irons, bending the latter into arches without much trouble."[36] Although Barnes, as usual, exaggerated, a photograph from the mid-1920s does show Tusko in his pen, with chains on his feet so he could be moved around for cleaning and other tasks, and the rails and iron buttresses of the pen are indeed bent (fig. 5.9). His tusks were broken off during these years in the pen. The story Barnes told was that one day an inspecting humane society officer arrived and went into the pen; Tusko did not appreciate the intrusion of an unwelcome guest, and when he charged, he shattered his tusks in a collision with the rails.[37]

Tusko was not always in his pen, though. He toured with the circus for most of the 1927 and 1928 seasons, as he did in the 1929 and 1930 seasons after the Barnes properties were sold to the American Circus Corporation and then passed on to John Ringling when he purchased the corporation.[38] It was in this period that Tusko became an entirely different kind of exhibit. Tusko was no longer the young elephant who had a handful of tricks, no longer the larger elephant with beautiful tusks leading parades of a wagon show and

Figure 5.9. Tusko in his pen, tusks gone. Circus World Museum, Baraboo, Wisconsin.

being taken to an arena in Mexico, no longer even the ride elephant of a man who wished to be seen as a king; Tusko became, instead, a spectacle of danger and chains. The chains made sense from risk- and elephant-management perspectives—there is little doubt that Tusko could be a difficult-to-handle and very dangerous elephant. But the chains also became a kind of exhibit in themselves. A head-on photograph of Tusko outside his pen in 1927 makes stunningly clear that although the chains were intended to restrain the animal, they would inevitably become a distinctive feature of this elephant (fig. 5.10).

Stories of Tusko "running amok" in Sedro-Woolley, popular tales of hunters facing off against the most dangerous game in Africa and India, and a whole series of stories about other elephants including Topsy in Coney Island in 1903, Hero in South Dakota, "Murderous" Mary in Tennessee in 1916, and Diamond in Texas in 1929, cultivated a desire in the public to see Tusko in chains (and perhaps the more chains the better).[39] The chains, often covered in rubber hoses in areas where they severely abraded the elephant's skin, hobbled Tusko so he could move only very slowly and prevented him from using his head, trunk, or legs to attack (fig. 5.11). Tusko came to embody both the spectacular and the dangerous qualities of the circus, two qualities that became quintessential to the American circus experience during the first half of the twentieth century. If the circus in the nineteenth century was a place of wonder and extraordinary feats, a place where visitors could see miraculous creatures like hippos or giraffes and be appalled, amazed, or fascinated by "freaks" and dancing women in the sideshow tents, with exhibits like Tusko in chains and new "death-defying" performances, the circus quickly evolved in the twentieth century into a new kind of place of wonder, a place where people might both fear and hope to see an elephant go mad or, perhaps one should just say it, even see someone killed.

This is, of course, the darker side of popular attendance at all kinds of obviously dangerous spectator events over the last century, and it is one of the only ways to make sense of the performance of Tusko in chains and how the circus promoted the exhibit (fig. 5.12). Tusko, as Walton puts it "was chained as no other exhibition elephant had ever been." The chains were the result of Tusko's efforts to exert his will in challenging circumstances, but they were also the outcome of a particular historical moment and way of thinking about elephants. For a whole series of reasons, we will not see this kind of exhibition of an elephant in the West again.

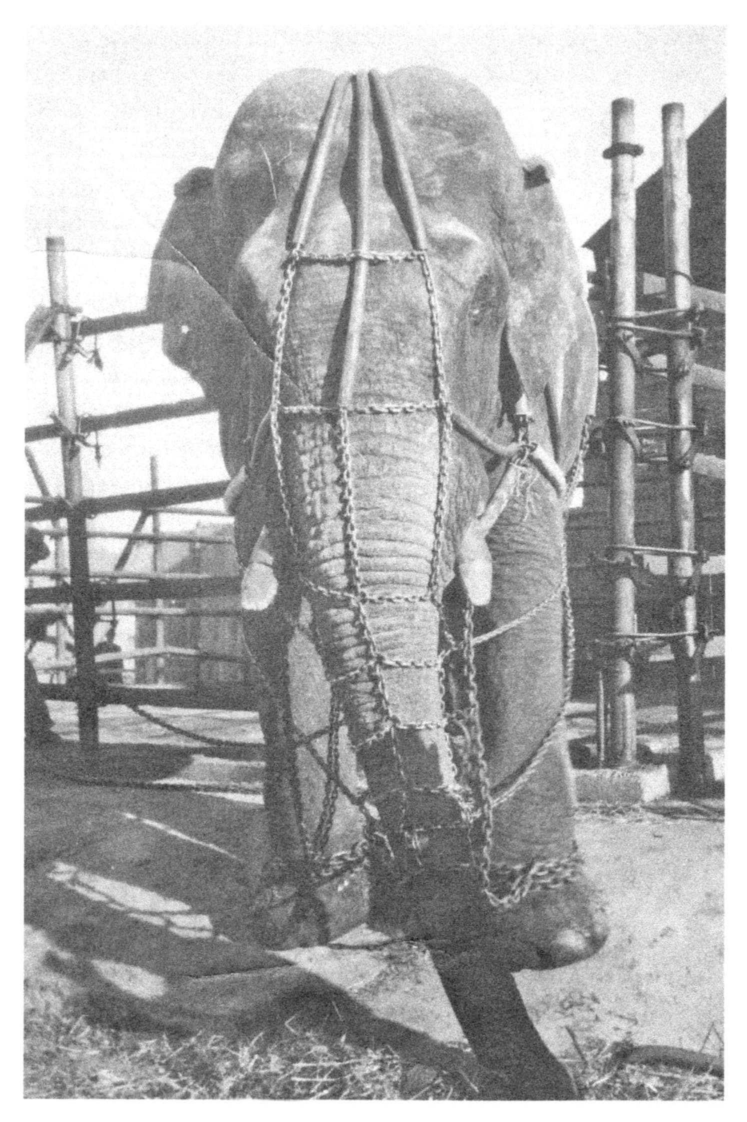

Figure 5.10. Tusko outside his pen in Baldwin Park, California. Circus World Museum, Baraboo, Wisconsin.

Removing the Chains

Tusko's life did not end, however, as an exhibit of danger and chains at a circus. His last season with the Barnes Circus was 1930. The year started at the end of March with a long series of stands in California, and in May and June the show was in Oregon and Washington. After stops in Idaho and Montana,

Figure 5.11. Tusko exits his railroad car. Circus World Museum, Baraboo, Wisconsin.

the show opened on June 18 in Alberta, Canada, and then started working east, arriving in Sudbury, Ontario on the first of July. By mid-July, the circus had reached New Brunswick, and at the end of the month and in the first days of August it was doing stands in Windsor, Digby, Yarmouth, Bridgewater, Halifax, and Truro in Nova Scotia. The show made its way back to New Brunswick, then stopped at towns in Maine, New Hampshire, Massachusetts, New York, Pennsylvania, Ohio, Indiana, Illinois, Missouri, Kansas, Oklahoma, Arkansas, Texas, New Mexico, and Arizona before finishing off the season in mid-October back in California.[40]

In early 1931, with the Great Depression deepening and circuses beginning to struggle, Tusko was sold to the first of a series of promotors and various others who thought they could make money charging admission to see the famous elephant. The first buyer was a man named Al Painter, described as "a dapper gentleman who wore spats and made a living promoting this and that." At the time, Painter was promoting "walkathons"—events in which contestants walked in circles night and day until only one was left standing. The one who won would then receive a cash prize, and Painter would make a windfall as well.[41] With the help of Jack O'Grady, one of Barnes's elephant

Figure 5.12. An exhibit in chains. Woodland Park Zoo Historical and Administrative Records, record series 8601-01, box 15, folder 1, Seattle Municipal Archives.

men, and Bayard "Sleepy" Gray, one of Barnes's teamsters, Painter exhibited Tusko in Portland and then Seattle at the Playland Amusement park for several months. Tusko was not traveling by rail anymore, and he was not walking between towns, either. The rest of his travels would be by truck, usually chained by the feet to a flatbed like a large tractor. After Seattle, Tusko was seen at the Yakima and Puyallup fairs and at the Hillsboro and Salem fairs in Oregon. Tusko was in a barn at the fairgrounds in Salem when Painter up and disappeared. The city took over ownership and then tried to sell Tusko at auction. After a failed effort that led to a top bid of $12, a boxing promotor named Harry Plant gained ownership at a second auction with a $200 bid—still a $100 below the city's reserve price, but the city took what it could get because it was obviously better to have $200 than an elephant.[42] Another elephant handler, George "Slim" Lewis, joined the O'Grady, Gray, and Plant in Salem, and they decided it was time to move on because "the dimes were coming in more and more slowly"[43] The group decided to go to Portland and they rented a building on the east side, near the center of the city on Water Street, which they got for $20 per month. Announcements appeared in papers across the country on November 29; one noted that Tusko, "against whose thick hide the barbs of fate have fallen fast in the past few months, had a new home."[44]

Audiences from far away continued to follow Tusko's adventures. Hopes were raised in Salt Lake City, for example, that Tusko could be purchased as a mate for the Salt Lake Zoological Society's Princess Alice. Later in the month, news spread that Tusko had a cold and had been given a giant toddy consisting of ten gallons of moonshine and a barrel of water that apparently helped his "surly manner."[45] Then on Christmas Day, 1931, according to newspapers, Tusko went on another "rampage." The whole business started the afternoon before when the men noticed that Tusko was fiddling with the pins holding the chains on his front legs. Normally, according to Lewis, when Tusko was going into musth the pins would be hammered flat, so the elephant could not unscrew them. This time, though, the keepers waited too long to do the hammering and Tusko had become unapproachable. Eventually, he managed to get the chains off his front legs. He then directed his attention to a wall of the building, which he battered down with his head. Policemen were alerted and stood outside with high-powered rifles, and then more policemen arrived with submachine guns ready. Tusko was still anchored, though, by long chains attached to his back legs. Hearing what was happening, the mayor sent word to the police not to shoot unless the elephant got his last leg chain off. Eventually, Lewis and O'Grady managed to get cables on Tusko's front legs by climbing through a cast iron water main that they had pushed near the elephant. The cables were hitched to trucks and Tusko was secured. Another round of articles appeared with titles like "Tusko on Rampage, Again: Dodges Firing Squad," "Big Elephant Rechained in Wild Battle," and "Huge Animal Saved from Firing Squad: Elephant Goes Berserk, but Mayor Steps in to Halt Guns and Quiet Beast."[46] The news, as it had always done, increased the dimes. In his memoirs, Lewis claimed that fifty thousand people visited Tusko over the next weeks, paying a dime a head. It seems an unlikely number, but even if ten thousand visited, that was still $1,000 in a time when money could be very difficult to find.[47]

In the spring of 1932, Tusko was in Woodland, Washington, and then the group moved on to Chehalis for about a month. Further stops were made in Olympia and Tacoma. Eventually, O'Grady, Gray, and Plant gave up because there was simply not enough money coming in, and Lewis was left alone caring for the elephant. For a period, it seems, the two stayed where they could using a tent or barns. For a while, Lewis recalled, they spent some time in a barn belonging to an old circus friend who had a place along Highway 99 between Seattle and Tacoma. It was there, Lewis claimed, that "Tusko spent the most pleasant days of his life" walking "at will in the fields and woods nearby."[48] At the end of the summer, Lewis and Tusko were back in Seattle for Fleet Week, and after that he was moved to a lot on 8th Avenue and Stewart Street. While he was there, with winter approaching and with few options left for Lewis, a meeting was called between Lewis, Mayor John Dore, Woodland Park Zoo director Gus Knudson, and Harry Ireland of the Humane Society. The group concluded that the best thing to do would be to declare Tusko a "public nuisance," seize ownership of the animal, and move him to the Seattle zoo. On October 8, Lewis chained a standing Tusko onto a flatbed trailer, and the elephant was moved to the Woodland Park Zoo in Seattle, his final home.

There had been, it seems, growing public dismay over Tusko's less-than ideal circumstances. While he was Salem, for example, a fund was apparently started by concerned citizens hoping to send him back to "Siam" where he would "no longer be subject to the ignominy of being valued at $12 on the auction block" and where he could be free from "triple-strength steel chains."[49] After the so-called rampage in Portland, an editorialist from the *Astoria Budget*, wondering whether Tusko would have been better off being shot down by the police instead of continuing his life of misery, opined that "the brain that could release heavy iron shackles by patiently unscrewing the nuts that bolt them to the elephantine legs, is quite capable of a sudden and terrible ennui against existence in a cold, dark, draft shack; occasional hangovers from a tub of bad moonshine; and the constant thralldom of heavy fetters."[50] The thoughts expressed in two handwritten letters from Rose Hellman to zoo director Knudson probably spoke for many. In the first, written the day after Tusko arrived at the zoo, Hellman expressed her relief that Tusko was finally in the superintendent's "good hands": "I *do* hope you will find a way to relieve him of the way he is *trussed up* in *chains*. It is cruel enough, the isolation from his own kind, the captivity, and the half-starved conditions he has had to endure, but to chain an Animal on all feet is rank cruelty. It would be better dead." In a follow-up letter on November 25, Hellman voiced her worry about what would happen if Tusko were sent back to the circus world: "I dread to see this great bulk of Flesh and Blood, this wise old man Tusko go into the hands of these cruel, Ignorant louts again. So much suffering can go on in that *huge Bulk*. It cannot *tell*. Do all in your power to hold him, bear with his tantrums. *Cruelty done it*."[51]

As for Lewis, he stayed on with Tusko, hired by Knudson as the elephant's keeper.[52] The zoo did face litigation from various quarters with claims of ownership, claims that appear to have led the director to learn as much as he could about the elephant's history, but the parks department, humane society, mayor, and Knudson seemed determined to keep Tusko at Woodland Park. They may have simply recognized that the elephant, obtained essentially without cost, would be an attraction at the zoo, and he certainly was, at least initially. Still, one must concede that the move to the zoo was likely an improvement for the elephant. Tusko ended up living in an actual building with heat, which had a yard where he could walk without being draped in chains. One should note, though, that most of the pictures of Tusko in the yard from that time show his front feet chained, standing on and secured to a wood board concealed from the camera with straw. In these photographs, it may look like he is just standing in the yard calmly, but he was actually unable to move. Still, it seems the zoo did not try to address Tusko's temper with moonshine, and there was even another elephant there, Wide Awake, who arrived in 1921. At the same time, it has to be acknowledged that he continued to be difficult to manage. While Lewis portrayed a mutually loving relationship with Tusko, there are records that make clear that Lewis had to consistently and brutally reinforce his dominance over Tusko in order to manage him. A report from the spring of 1933 detailing efforts to pull Tusko with two trucks to a side of his stall noted that Tusko resisted by "rearing, plunging, and pulling on the chain with the fastened foot, but the trucks held the strain. He tried kicking, hitting with his tail and every known means to strike Slim who was busy keeping out of the way and punishing him with the hook. Finally Tusko acknowledged defeat by ceasing to fight. During the evening of the same day training was again resumed which ended in submission of Tusko." The report concluded, "Tusko is not reliable or subdued by any means but only for a short time as control must be drilled in to him from time to time."[53]

Seven months after he arrived at the zoo, on Friday June 9, 1933 around noon, Tusko collapsed while out in the yard of his enclosure. He got back to his feet. At 3 p.m. he was down again. He was given fifty sodium salicylate tablets as a pain reliever. At 6 p.m. he showed weakness in his hind quarters. At 11 p.m. he was leaning against a wall but appeared to have difficulty controlling his muscles, he wheeled and collapsed again. Lewis began removing Tusko's chains. In his official report of the events, Lewis wrote that by the time the director arrived at just after 11 p.m., "I had all the chains off and took off the anchor bracelet from around his right hind foot. That left him entirely without chains. Every few minutes he made an effort to get up but his hind feet wouldn't hold him up. He had good control of his front feet."[54] Lewis and others stayed with Tusko through the night. Shortly before 10 a.m. the next morning, June 10, 1933, Tusko died. He had been out of his chains for about the last eleven hours of his life. A large blood clot was found in his heart. He was about thirty-five years old; most people seemed to think of him as old, but he was still a relatively young adult.

Circuses and Zoos

In the spring of 2003 I participated in a three-day workshop on elephants and ethics held at what was then called the Conservation and Research Center and is now known as the Smithsonian Conservation Biology Institute in Front Royal, Virginia. The event brought together field scientists, zoo professionals, animal rights activists, representatives of the circus industry, and academics to talk about elephants. The discussions were collected in a volume titled *Elephants and Ethics: Toward a Morality of Coexistence*, edited by Christen Wemmer and Catherine A. Christen, the workshop organizers.[55] A key element of the workshop was that after each day the participants would gather in groups and try to reach agreements regarding general statements about human-elephant interactions that had come up over the course of the day's presentations. It was notable that when the workshop was over and the final list of consensus statements was compiled, the participants had not agreed on anything regarding elephants in circuses. When it came to zoos, the participants could agree, for example, that not every zoo has high-quality facilities for keeping elephants, that not every zoo should expect to be able to keep elephants, and that conservation efforts for the species should focus on protecting elephants in their natural habitats and not in zoos. When it came to circuses, though, the lines were drawn so quickly and resolutely between those who felt that circuses were the worst possible places for the animals and those who felt that elephants could flourish in circuses that any kind of compromise or agreement was simply impossible.

According to the first group, circuses, elephant rides, and other venues where elephants perform or otherwise entertain the public were simply self-evidently cruel and exploitative. For these critics, the only acceptable outcome of a discussion about the future of elephants performing in circuses would have been an agreement calling for the banning or criminalization of keeping elephants solely for purposes of human entertainment. The second group consisted of people who had devoted their lives to being with elephants, who believed they took excellent care of the animals, and who had been vilified by their critics for decades. A Venn diagram of circles representing the diverse interests in the room would have shown that the circles occupied by representatives of the circus industry and the animal rights activists did not overlap in any meaningful way. And this was a perfectly acceptable outcome as far as both of these groups were concerned.

The inability to reach any kind of consensus about the circus was probably inevitable and should perhaps have even been anticipated. For their part, the organizers, a conservation biologist and an environmental historian, had been able to find common ground with everyone in the room. I think that from their perspective, the goal was to build understanding and break down animosities by focusing on improving the circumstances of elephants. I found myself in a similar position as the organizers. I had come to know people over the years from both the zoo and circus industries and the animal rights community, and I had found myself listening carefully and with admiration to people with very different views about elephants and the history of human involvement with them. There have been times over the years when I have been frustrated with activists because yelling into a megaphone tends neither to bring people to the table for a thoughtful discussion nor to convey the subtleties of a particular situation in which an elephant may find herself. The ups and downs of life of any individual animal, its days of frustration and happiness, are not, in my view, captured well in slogans on billboards and soundbites on television, in Facebook posts, or through online petitions. However, I have also been frustrated by people connected to circuses and zoos who too easily dismiss any critic of their practices as a radical extremist and who refuse to accept that critiques over the last century have, in fact, improved circumstances for animals, if not necessarily for the particular animal being fought over. Even though the free Keiko effort, for example, resulted in the whale's death, because it turned out that he was unable to survive on his own in the wild after having been in human captivity, sensitizing the public to the challenges faced by captive whales has improved the circumstances of some captive orcas and slowed the practice of pulling more whales into captivity from the wild.[56]

With the 2015 decision of the Ringling Brothers and Barnum & Bailey Circus to phase out elephant performances in its shows, a step that seemed to lead to the whole circus shutting down in 2017, it may appear that the days of using elephants simply for entertainment in the United States are numbered. However, the ancient history of pressing elephants into service as "spectacles," as ways to claim that whatever one is doing is bigger, more important, and more impressive than anything done before, makes clear that as long as these animals continue to exist, they will continue to be used to get people's attention.[57] Elephants, of course, still travel with smaller circuses, are hired out for special events and movies, and provide rides at various kinds of fairs all over the United States, and these types of operations will likely continue for the foreseeable future. Elephants are used in other venues, though. Indeed, if elephants were never to be used again to provide rides or other forms of entertainment, it seems likely that they would still be used to make arguments about a wide array of other issues. Even if when people sit down before a "jumbo" serving of French fries or look forward to a "mammoth" sales event at a car dealership, they do not consciously think about the connections between these words and the spectacle of elephants, this spectacle is far from gone from our experience.

William Hornaday wanted an adult, male, Asian elephant, with tusks because having such an animal (along with other elephants, rhinos, hippos, and tapirs) in a spectacular elephant house would make the argument that the New York Zoological Park was the most impressive in the world. When people went to the Bronx Zoo to see Gunda chained to a wall, they clearly did not go for Hornaday's vaunted reasons of science or education or because they were concerned about the conservation of elephants; they went see the largest and by some accounts most dangerous animal in North America straining against its chains. They went to see a show. People went to the elephant house to be entertained, and when the elephant did not meet their expectations, they goaded him with taunts, calling his name and pretending to throw him food. There is an ugliness in all of this, an ugliness that was as apparent and undoubtedly as exasperating to Gunda's keepers and the administration of the zoo as it might be to many today looking back at Gunda's life. But the demand of the public to be amused at the zoo and the circus points to what has always been one of the core reasons for zoological collections. As much as Rudolf II's remarkable menagerie in Prague in late sixteenth century was part of his larger ambitions for the Habsburg court, his desire to make it home to the most noteworthy sights and minds of the time, his famous menagerie and its animals were also, of course, highly entertaining for those who visited. Similarly, even though the collections of the Zoological Society of London, which opened in 1828, were originally intended to be for the exclusive scientific use of the fellows of the society and their guests, the gardens proved to be more important as a destination for socializing than for science. Before long, the London Zoo was opened to anyone who could buy a ticket, because, as the 1869 song "Walking in the Zoo" has it, "The okay thing on Sunday is walking in the zoo." For Rudolf II in the sixteenth century, the London Zoo in the nineteenth century, and the Bronx Zoo in the twentieth, the presence of elephants was a critical part of what made the collections self-evidently entertaining or "okay."

Still, as much as entertainment has been central to the history of zoos, most people nevertheless draw a distinction between zoos on the one hand and circuses and other venues where animals are used for entertainment on the other. While it may be plain enough that people go to zoos primarily because they think they are "fun," zoos, or at least what are often called "good zoos" or "better zoos," tend to be regarded along with various kinds of museums as broadly educational. Some zoos have succeeded in distancing themselves even further from pure entertainment by convincingly arguing that they are part of larger conservation efforts. Many people clearly believe that these better zoos can both help the public understand the urgent need for conservation and serve as sanctuaries and breeding centers for species that are threatened or have even become extinct in the wild. Certainly, as the story of Gunda makes clear, the distinction between zoos and other kinds of entertainment can be fuzzy, but the intuitive distinction that most people make is not just the result of misunderstanding the facts.

In the nineteenth century and through most of the twentieth century, the lives of elephants in zoos and circuses were clearly not that different. The techniques used to handle the animals were essentially the same, most of the animals were trained to perform tricks for the pubic, and the animals were overwhelmingly seen by the public (and the proprietors) as objects of entertainment in both venues. More than anything else, what distinguished the lives of elephants in the circus was the travel, and that aspect might not necessarily have been bad for the animals physically, intellectually, or emotionally. Tusko's eventual move to a zoo at the end of his life, moreover, was far from unusual. During winter seasons, circus elephants were often placed in zoos, and elephants that had become difficult to handle in circuses or needed places to stay for one reason or another frequently found themselves transferred or sold to zoos where they sometimes lived out their days. From the visitors' perspective, too, walking through the menagerie tent at the circus—which often had quite large collections of unusual animals—was really not that different, logistically speaking, from visiting a small zoo in the period. Carnivores and various other animals were to be seen in cages, other animals were in stalls, and quite a few animals could be fed and petted.

In short, the line dividing zoos from circuses throughout the nineteenth and first half of the twentieth century was not clearly drawn. But zoos and circuses were different. Zoos in this period were often seen as calm, clean, orderly, thoughtful places, where educated people could stroll and see animals in decidedly controlled spaces—the sort of place imagined by Hornaday—while circuses were always seen as more exciting, as places of adventure where there was, one hoped, at least a whiff of danger—if only for the performers. That element of danger became, if anything, more pronounced in the twentieth century, as did the demand for increasingly spectacular performances. In both the worlds of zoos and circuses, one of the key drivers of change has been the constant effort to outdo the competition. For zoos, that has tended to mean that every zoo has wanted to make itself stand out with larger buildings, larger exhibits, rarer animals, and more intense "personal" experiences; for circuses, it has meant ever more spectacular kinds of performance with at least the appearance of greater risk and danger.

Tusko's life in the American circus spanned a period of critical change in the lives of elephants in circuses, a period that began with relatively uncomplicated and amusing tricks by a couple of elephants performed before audiences in rural towns to huge shows traveling by rail to larger cities, shows that traded on hyperbole, peril, and the demand for ever more spectacular entertainment. For Tusko this meant that in just over a couple of decades he went from being a young elephant with a handful of tricks to a chained monster to, finally, an unwanted "old" elephant who was lucky to find a home in a Seattle zoo. None of these descriptions present a complete or accurate report of his life, but they do help bring to light a history of circuses and how elephants came to play such a critical role in that history.

The Last of Its Kind

At the end of a row of elephant skulls at the storage facility in London rests the trophy of a gigantic bull elephant. It is an arresting and unsettling sight. The specimen is unusual at first glance owing not only to its tremendous size but also to a custom iron armature supporting the mandible and the dark color of the bones, the legacy of a long-ago varnish that was intended, it seems, to make the skull a more appealing decoration for a home (fig. 6.1). The tag on the skull notes that the specimen is from the no-longer-recognized subspecies *Elephas maximus bengalensis* and was collected in 1888 in the Garo Hills in Assam by G. P. Sanderson. Again, the bones sent me to libraries and archives and soon a richer history began to come into focus.

By the time of his death in 1902, George Peress Sanderson had become one of the names indelibly connected to the history of the British imperial rule in India, partly because he appears to have been the basis for Rudyard Kipling's character Peterson Sahib in his short story "Toomai of the Elephants"—a story that has also been made into multiple films. In 1864, when he was sixteen years old, Sanderson, who was born in India to missionary parents but sent to England for school, arrived in Mysore, learned the local language of Kannada, and was eventually appointed to oversee part of an ancient canal system. "Owing," he writes, "to the advancement of officers above" him, he was eventually put in charge of the whole system of some 716 miles of canals in 1868. The appointment, Sanderson argues, was ideal for a young man more interested in hunting than civil service. As he writes in his 1878 memoir *Thirteen Years among the Wild Beasts of India,* "I had a large extent of country, including several fine jungles . . . to travel over in the prosecution of my work. I had a sufficient salary to afford a good battery, and the money necessary for getting good sport; and I spent most of my leave and all my cash upon it."[1] In 1873 he

Figure 6.1. Skull of an elephant shot by G. P. Sanderson in 1888. Photograph by Helen J. Bullard, with permission of the Natural History Museum, London.

was given permission to try to capture several herds of elephants in the forests near Mysore. His efforts were successful, and in 1875 he was temporarily put in charge of the Bengal Elephant-Catching Establishment, working in the Garo and Chittagong hill tracts. After capturing eighty-five elephants in Chittagong, Sanderson returned to Mysore and was put in charge of elephant operations there, using a corral system similar to the one used in Ceylon and described by Tennent.

Thirteen Years among the Wild Beasts of India was written very much in the spirit of Samuel White Baker.[2] Sanderson's description of an elephant's charge, for example, could just as easily have come from Baker's pen: "The wild elephant's attack is one of the noblest sights of the chase. A grander animated object than a wild elephant in full charge can hardly be imagined." Calling the tiger's charge "an undignified display of arms, legs, and spluttering," Sanderson compares the charge of a large tusker to the "rapid advance of an engine on a line of rail."[3] Hunting after almost anything for Sanderson and Baker, as it would be for Theodore Roosevelt as well, was thrilling, civilized, and "manly," but pursuing elephants presented an extreme version of the hunt, a version that seemed almost inherently ennobling. For these men, tracking and then standing up to a charging elephant before sending the final shot was a test of courage, of huntsmanship and bushcraft. In the imperial contexts in which these men were active, moreover, hunting an elephant was also an ideal expression of their prerogatives. Hunting for the pot, hunting for commercial reasons, or encouraging natives to hunt was almost always frowned on by this type of hunter. But hunting alone (or what they called "alone," which typically meant accompanied by native gun bearers, carriers, and personal "boys") was always seen as a high form of "civilized" activity.

Even when confronted with irrefutable facts about the disappearance of game, these men almost reflexively pointed to others as the problem and were convinced that through the creation of reserves and the requirement of permits, through banning hunting for commercial purposes, and through forbidding indigenous peoples from hunting, the hunt could be preserved for those motivated by higher purposes such as the pursuit of knowledge or the desire for a personal challenge. In his memoir, for example, Sanderson notably pauses before his chapter on elephant hunting, what he called "the grandest of all field-sports," to wonder whether it makes sense to write about the subject at all because hunting elephants had been banned throughout India and Ceylon to protect remaining stocks of wild herds.[4] But having stopped to consider this state of affairs, Sanderson perishes the thought, insisting that the hunt would inevitably return—a belief that proved to be true. Without knowing it, though, Sanderson was writing at a moment when hunting, and especially hunting elephants, was just beginning to be subjected to a wider critique. Around the end of the nineteenth and beginning of the twentieth centuries, there were other writers, including hunters, who began to wonder not only whether hunting should continue to be permitted, but whether elephants even had a future. If sport hunters like Baker, Sanderson, and Roosevelt were able to convince themselves that they were not the problem, others began to contemplate what they saw as the inevitable extinction—the disappearance—of elephants.

The Spell of the Elelescho

> The traveler probably sees only a tree-like shrub. It covers many a ridge and the lonely plains of the steppes and sends into the distance its spicy perfume.
>
> —Carl Georg Schillings, *Der Zauber des Elelescho* (1906)

Near the end of 1904, the German hunter and explorer Carl Georg Schillings published a groundbreaking book with the title *Mit Blitzlicht und Büchse: Neue*

Beobachtungen und Erlebnisse in der Wildnis inmitten der Tierwelt Äquatorial-Ostafrika, translated by Frederic Whyte into English in 1906 with the title *With Flashlight and Rifle: A Record of Hunting Adventures and of Studies in Wild Life in Equatorial East Africa.*[5] *With Flashlight and Rifle* became an almost instant classic, thanks partly to promotion of it by Theodore Roosevelt. In his *Outdoor Pastimes of an American Hunter* of 1905 Roosevelt urges that *Mit Blitzlicht und Büchse,* which he calls the "best recent book on the wilderness," be immediately translated into English. Schillings, Roosevelt claims, is "a great field naturalist, a trained scientific observer, as well as a mighty hunter," and he concludes that "no mere hunter" could ever contribute as much to knowledge as Schillings does in this book. He argues that every modern big game hunter should be like Schillings: an "adventure-loving field naturalist and observer."[6] Roosevelt was deeply impressed with Schillings' photographs, and the plates in his own *African Game Trails* of several years later show the influence of the German work. But the writing too earned his praise. As he noted once in discussing books on hunting, "the most exciting events, if chronicled simply as 'shot three rhinos and two buffalo; the first rhino and both buffalo charged,' become about as thrilling as a paragraph in Baedeker"—the popular travel guides of the time. "Really good game books," Roosevelt insisted, "are sure to contain descriptions which linger in the mind just like one's pet passages in any other good book."[7] Schillings, Roosevelt felt, wrote a good book. The general reception in Germany, England, and France seems to have echoed Roosevelt's enthusiasm as word spread about the book.[8] Roosevelt wrote to William Hornaday in January 1905: "By George, what an amount of work remains to be done as regards the natural history of big game, and how little big-game hunters have accomplished in this direction! You know Schilling's book, do you not!"[9]

At first glance, what made *With Flashlight and Rifle* such an unusual book was the over three hundred photographs that were taken using what was then cutting-edge field technology, including telephotography and new methods for night photography. Unlike previous works by other authors that primarily reproduced photographs of already dead animals, Schillings tried to capture scenes of living animals and their behavior. He wanted to photograph animals, people, and landscapes in action.[10] As a body of work, the photographs broke new ground and were described as "Natururkunde"—documents of nature—by the director of the Berlin Zoological Gardens in his introduction to the German edition.

The photographs are remarkable. Schillings' text, however, is also often not what might have be anticipated from a book with an image of a downed elephant on the cover.[11] With the opening chapter, titled "The Tragedy of Civilisation," Schillings makes clear that the book is more than simply stories of hunting adventures. The underlying message is that the spread of civilization was bringing an end to so much life on the planet. "Far from the smoky centres of civilisation," Schillings writes, "with their rush and turmoil and the unceasing throb and rattle of their machinery, there is at this present moment being enacted a grave and moving and unique tragedy." Schillings explains that as explorers move ever deeper into Africa and other areas around the globe, the original inhabitants are pushed out, and with them disappear "a rich and splendid fauna, which for thousands of years has made existence possible for the natives." The explorers and colonists kill off any animal from which a profit can be made and kill off others simply because they are in some way inconvenient. "Never before in the history of the world," he states, "have whole hordes of animals—the larger and stronger animals especially—been killed off so speedily by man." It is not, moreover, just the indigenous peoples and animals that disappear. Trees and forests are also displaced by invading plants that follow in the wake of the Europeans. In the end, he observes, the colonist "will destroy everything that is useless to him or in his way and will seek only to have such fauna and flora as answer to his needs or his tastes."[12]

There is repeatedly an urgency in *With Flashlight and Rifle*, a sense that the time for especially big game was quickly running out. Schillings points, for example, to the disappearance of the American bison. Noting that just a few decades earlier, millions of bison had "roamed over their wide prairies," he laments that they have now "gone the same way as the vanished Indian tribes that once lived side by side with them." He continues, "Soon a long list of other noble specimens of the American fauna will follow them," and he praises President Roosevelt's efforts "to stave off this inevitable calamity."[13]

Compared to other popular hunting books of the time about Africa, *With Flashlight and Rifle* was quite different owing its concern over the disappearance of game. It was still, however, recognizable as a hunting memoir, organized in the usual way with chapters devoted to hunting different species—elephants, rhinoceroses, hippos, buffaloes, crocodiles, giraffes, zebras, lions, and so forth. There are the predictable hair-raising stalks and last-minute shots at this or that wild beast. Schillings was, after all, in Africa to hunt and to preserve specimens, and he returned to Europe from his two expeditions with massive collections. While Schillings at one point criticizes missionaries in Tanganyika for killing thirty-seven lions over four years, mostly through the use of strychnine, he all by himself returned to Germany with "some forty specimens of lions, about thirty-five leopards, as well as large numbers of hyenas, jackals, and other beasts of prey."[14] He also brought a bird collection of over one thousand specimens. Paul Matschie, curator of the Royal Zoological Museum in Berlin, said he had "collected a greater number of different species than any other traveller before him."[15] This was a book that a hunter like Roosevelt could admire; it was also a book that a conservationist like Roosevelt could admire. Alarmed by the loss of the extraordinary animals of the world, the book urged hunters to take the lead in protecting game and to collect specimens for science before all the animals succumbed to what seemed eventual, inevitable destruction. The slaughter perpetrated by farmers, colonists, native peoples, pot hunters, and market hunters had to be stopped or at least slowed, Schillings argued. While progress and civilization would, from his point of view, unavoidably prove incompatible with wildlife, Schillings hoped that thoughtful hunters and thoughtful governments could possibly protect at least a few fragments of land on which the remnants of a past glory of wildlife could be sustained in game reserves.

With Flashlight and Rifle was an important milestone in the emerging conservation movement at the beginning of the twentieth century. However, a second memoir by Schillings published two years later in 1906, with an English translation in 1907, went much further.[16] *Der Zauber des Elelescho* (literally, "the spell of the elelescho") may have been, as the historian Bernhard Gissibl has argued, simply an effort to capitalize on the success of *With Flashlight and Rifle*, but it also has a quite different feel about it. The difference is evident even on the cover, where the poetic words that make up its title are positioned below a striking image of a living giraffe lying quietly in the grass (fig. 6.2). From the first sentence, it is clear that this book will not simply be a repeat of *With Flashlight and Rifle*. The English translation, *In Wildest Africa*, begins: "On the afternoon of January 14, 1897, a small caravan of native bearers, some fifty strong, was wearily making its way across the wide plain towards its long-wished-for goal."[17] The party had been marching from Lake Victoria, where Schillings had fallen gravely ill with fever, to the higher elevations around Lake Nakuru, which was at last coming into sight. The banks of the salty lake in that season were spread with new pastures, and Schillings breathed fresher air while looking out on "thousands and thousands" of little Thomson

Figure 6.2. Cover of C. G. Schillings, *Der Zauber des Elelescho* (Leipzig: R. Voigtländer, 1906).

gazelles feeding on "the fresh, green, grassy meadows of the lake margin, or scattered over its pebble beds of obsidian, augite, and pumice-stone."[18]

Thinking back on his days by the lake, Schillings recalls, "Out of the many memories of those days, that still work on me like magic, there is one above all that has a special meaning for me: Elelescho!"[19] For Schillings, the sight and scent of the silver-grey-leafed wild camphor bush (*Tarchonanthus camphoratus*), still known locally as "leleshwa," was "linked with the plunge into uninhabited solitudes, with self-liberation from the pressure of the civilisation of modern men and all its haste and hurry."[20] Schillings imagines that while the plant was undoubtedly still to be found there, the old magical properties of the scent had likely been destroyed by the advance of European civilization, by the railroad and all that came with it. Remembering a time when the "spell of the wilderness" and the "mood of the night" "ached on the senses like the nocturne of some great tone-poet," Schillings mourns the passing of an older time.[21]

Tone poems would be going out of style by the 1920s, but when Schillings was writing this book, they were still a popular form of orchestral composition and had been for over three quarters of a century. Constructed as a single movement, tone poems linked musical themes to narratives in dramatic paintings, famous stories, or legends, and listeners were encouraged to think of the tales while they listened to the music.[22] Schillings' brother, Max, to whom the memoir is dedicated, was a composer of tone poems, and it seems clear that Schillings intended his book to have a lyrical or musical quality. While the book is still very much on the ground in Africa, there are many moments that seem intended to resonate on a more aesthetic level. In the opening chapter, for example, there is an extended section describing a dream, a passage that would never have found a home in *With Flashlight and Rifle*. Schillings writes in a way that would be both thematically and aesthetically echoed by Roosevelt a few years later:

> I seemed to see before my eyes what happened here in primeval times—how volcanic forces, strange, boundless, and terrible, had built up and given form to the country around me here, destroying all living things, and yet at the same time preparing the conditions for the hotly pulsating waves of life of later days. In my mind I saw pass before me wondrous mighty forms of the animal world of the past, long since extinct.

The dream continued as a herd of elephants, hundreds strong, led by its "aged leader, a female elephant of exceptional size" came down to the shores of the lake to roll in the mud and drink. Among the large forms of the great beasts, there were also little elephants, just weeks old, and older ones who tended them, protecting them from harm. He then saw "herds of giraffes, hundreds strong, come down to the lake." Then, along with all kinds of antelopes, he saw "innumerable buffaloes" "coming for a refreshing bath."[23] In his dream, Schillings is able to move among the animals, and they show no fear. He eventually meets up with a caravan of Arabic traders and falls asleep listening to one of the men talk about the ivory trade. When he wakes, he realizes that all he had seen had been a dream, a dream brought on by the "spell of the Elelescho"—the fragrance of the plant had made it possible for him to witness Africa in a distant past. "In a hundred years from now," he concludes, "wide regions of what once was Darkest Africa will have been more or less civilised, and all that delightful animal world, which to-day still lives its life there, will have succumbed to the might of civilised man." In that future time, he notes, those who owned the horns, hides, tusks, skulls, and other specimens of the lost African fauna, would be selling their collections for their weight in gold. "No one will be able to understand then," he states, how it was possible that all was lost to "the interests of trade, and to the recklessness of the new settlers in those lands."[24]

Throughout *In Wildest Africa*, Schillings aims to make his readers *feel* the destruction of the wildlife, the cultures, and the land of Africa as much as he attempts to convince them to intellectually accept that it was happening. There would come a time, he was convinced, when most of Africa's large animals would be extinct. The most significant loss among these, of course, would be the elephant. The problem, from Schillings' perspective, was simple: commerce. Because people were willing to pay for ivory to make billiard balls, traders were driven to try to acquire as much as possible.[25] As a result, "powder and shot are at work day and night in the Dark Continent." It was not the European hunters, though, that were primarily responsible for most of the destruction. Rather, he thought, it was the native peoples who killed most of the animals to feed the European markets. Echoing other hunters, he insisted that unless the governments of Europe acted to stop the hunting of elephants by native peoples, "in a very short time the elephant will only be found in the most inaccessible and unhealthy districts." The end was inevitable, he argued, and whether it came in "thirty or forty or fifty years," those were nothing compared to "the endless ages that have gone to the evolution of these wonderful animals."[26] "Once again," he writes, "civilisation will have done away with an entire species in the course of a single century."[27]

For Schillings, elephants were threatened with extinction not because of sport hunters seeking trophies but because the desire for ivory led all kinds of people to kill every elephant they could find. While he did not consider sport hunting problematic, he did believe in hunting with a higher purpose—namely, to collect scientific specimens and to build collections of animals that would serve as records of the past when the animals eventually disappeared from the wild.[28] Two very different versions of a particular story that is related in both of the memoirs of photographing two bull elephants highlight Schillings' connection to the longer history of hunting elephants, on the one hand, and more recent sensibilities about the animals, on the other. In the version of the story in his first book, *With Flashlight and Rifle*, Schillings relates that he would never forget the time that he had waited "in vain for sunshine" for days on a hill, hoping to get some good photographs of elephants. "As soon as I had succeeded in this," he recalls, "the moment seemed at last to have come when I might kill the two bull elephants in question."[29] Schillings grabbed his guns and tore after the two elephants, but despite his efforts the animals got away. Two days later, though, he came on them again. Once more the hunt started and after over an hour pursuing the bulls, Schillings spotted them emerging from a mud wallow in a deep ravine and moving into a dense thicket. "It was heart-breaking!" he exclaims. "One instant sooner and both elephants would have been lying dead in the mud. Animals with tusks weighing two hundred pounds! Elephants such as have hardly fallen to any European hunter in the whole length and breadth of Africa!"[30] He followed the animals into the thicket, but they escaped.

The version of the story in the second memoir is notably different. Schillings writes that he had been wanting to take photographs of elephants moving freely in the wild. With the limitations of his technology, though, the shot was going to be difficult, but he felt that he might have a chance from a hill that overlooked a valley with a watering hole frequented by the animals.[31] To get from his camp to top of the hill, he explains, required five hours of hiking every morning, mostly in the dark; when he finally arrived, he was cold and "drenched to the skin" from his exposure to wet grasses and bushes, but he would not start a fire for fear of spooking the animals. Schillings then spent weeks hoping for a shot with the telephoto lens of his camera.[32] Finally, one day the clouds cleared and he spotted two very large elephants moving in the valley in the company of a giraffe (fig. 6.3). "I shall never forget," he writes, how "the mighty white tusks of two bull-elephants shone out in the hollow so dazzlingly white that one must have beheld them to understand their extraordinary effect."[33] Admitting that while he was taking the photographs, he wanted to shoot at least one of the large elephants with his guns, he nevertheless decided against it. He writes: "I had a hard struggle with myself. But the wish to secure the photographs triumphed. No museum in the world had ever had such a picture. That thought was conclusive."[34] The second memoir includes four full-page photographs of the bulls, taken at a distance of over four hundred yards, but Schillings does not indicate that he spent the following days trying to kill the two bulls. Reflecting on the images and this key experience in

Figure 6.3. Two elephants and a giraffe, in C. G. Schillings, *Der Zauber des Elelescho* (Leipzig: R. Voigtländer, 1906).

the second memoir, Schillings concludes, "these two, long since killed, no doubt, will continue to live on in my pictures for many a year to come."[35] The photographs, of course, were also trophies, but how Schillings told their story changed radically. In *With Flashlight and Rifle*, Schillings makes it clear he definitely wanted the photographs, but it was just as clear that they were not his priority, his priority was rather the hundreds of pounds of ivory. In the second account, published two years later, the whole story of trying to kill the bulls disappears, and the remarkable photographs become the focus of the text.

And this brings us back to the cover of the German edition of the second memoir, the image of a giraffe resting calmly in the grass. The image comes from a chapter devoted to giraffes entitled "A Vanishing Feature of the Velt" and appears on a page that precedes the concluding paragraphs of the chapter (fig. 6.4). "With sad, melancholy, wondering eyes," Schillings closes the chapter, "the giraffe seems to peer into the world of the present, where there is room for it no longer. Whoever has seen the expression in those eyes, an expression which has been immortalized by poets in song and ballad for thousands of years, will not easily forget it." "The day cannot be far distant," he concludes, "when the beautiful eyes of the last" giraffe "will close for ever in the desert."[36] The photograph is striking not least because unlike all the other photographs of giraffes in the book, this one has been taken at exceedingly close range. Remembering the cumbersome photographic technology of the day, we are forced to accept that this image—the image that also graces the cover of the book—was taken after the giraffe had been shot and crippled. This

Figure 6.4. Giraffe, in C. G. Schillings, *Der Zauber des Elelescho* (Leipzig: R. Voigtländer, 1906).

giraffe is *not* lying peacefully in the grass; the photograph was possible only because the animal was not able to get up and run away. This is not a photograph of a quiet existence on the veldt, an existence without the threats of Western hunters and their guns. This is a photograph of a dying giraffe. In the end and despite Schillings' own claims, this photograph, like so many of the hunter's "documents of nature," is about dying and death and even about extinction itself.

The Alarm

> Men have spoken of darkest Africa, but the dark chapters of African history are only now being written by the inroads of civilization.
>
> —Carl E. Akeley, *In Brightest Africa* (1920)

Schillings was far from a lone voice describing what we now usually call the sixth great extinction. In fact, awareness of the disappearance of wildlife around the planet can be found going back at least to the middle of the nineteenth century and even the late eighteenth century. A critical moment was undoubtedly when Cuvier declared in 1796 that the mammoth was an extinct species, after which the identification of many other extinct kinds of animals quickly followed, including varieties of rhinoceros, hippopotamus, bear, lion, tiger, hyena, the giant ground sloth, and the Irish elk.[37] Then in the 1820s and 1830s, the fossils of the first dinosaurs, the Megalosaurus and the Iguanodon, were discovered in England. Nevertheless, even while people were beginning to accept that formerly existing species were no longer living on the planet, the question about just how a whole species might disappear was still hotly debated. Some people, including Cuvier, believed that the only acceptable explanation for the destruction of species was a catastrophic event (or series of events) like a biblical flood. Others pointed to the ideas of the English geologist Charles Lyell. In his *Principles of Geology* (1830–33), Lyell argues that the world was immensely older than had been popularly imagined and that the geology, temperatures, and environments of the planet had changed over time. According to Lyell, species naturally moved into areas where they did not formerly exist and there often found easy prey that they extirpated, and he felt that the history of the planet was marked more by gradual, uniform change rather than one or more cataclysms. Thinking about discoveries of extinct forms of quadrupeds but to an even greater extent about different kinds of sea shells in the fossil record, Lyell argues that species disappeared through a great variety of causes. One of these causes was man. "If we reflect," Lyell states, "that many millions of square miles of the most fertile land" has been "brought under the dominion of man," we must accept that the "annihilation of a multitude of species has already been effected." Moreover, as "the colonies of highly civilized nations spread themselves over unoccupied lands," many more species would naturally disappear.[38] Lyell's arguments were convincing to many, but others, especially those committed to Christian teachings, remained steadfast in their beliefs about the perfection and stability of creation. Despite the debate over causes, though, most reputable naturalists by the middle of the nineteenth century agreed that species had disappeared in the past and that they were disappearing in the present as well.

By the time that Schillings wrote his chapter titled "A Dying Race of Giants" on the destruction of elephants in the second memoir, naturalists had been actively discussing extinction for a century. As Schillings took his telephoto images of two of the last giant tuskers—animals, he thought, almost from another time—he no doubt felt he was witnessing the final days of an extinction playing out before him. When Lyell wrote his *Principles of Geology,* his most dramatic example of a modern extinction was the dodo, a bird that had been wiped out in the early seventeenth century just decades after its discovery on the island of Mauritius. The blue antelope disappeared forever around 1800. Steller's sea cow, discovered by Europeans in 1741, was likely also gone by 1800, and the South African quagga disappeared by the 1880s. By the beginning of the twentieth century, the passenger pigeon was extinct in the wild, the American bison was all but completely annihilated, and large game was disappearing from whole swathes of Africa.

With the destruction of millions of birds for the millinery trade and for food, the dramatic collapses of various species sought by fishing fleets, and the decline in fur seals, sea otters, and several species of whales, Schillings and many of his contemporaries had reason to believe they were witnessing the end of much of the natural world. In 1910 William Hornaday notes that given "the inexorable disappearance of the grand game animals of the world," there was one vital task: that of "gathering now the collections that will adequately represent" the animals hereafter.[39] He concludes: "It is indeed time for the men of to-day who care for the interests of the men of to-morrow, to be up and doing in the forming of collections that a hundred years hence will justly and adequately represent the vanishing wild life of the world."[40]

One of those who apparently cared for "the interests of the men of tomorrow" was Carl Akeley, Hornaday's acquaintance at the American Museum of Natural History in New York and the man who came out to the Bronx to shoot Gunda and also Congo. By the time Schillings' second memoir was published in Germany, Akeley was already in the midst of his second expedition to Africa to collect animals, and he, like Schillings, had become convinced that he had limited time to do so. Akeley had begun his career as a taxidermist at Ward's Natural Science Establishment in Rochester, New York, then worked at the Milwaukee Public Museum in the late 1880s and 1890s before moving to the Field Museum in Chicago in 1896. After this move, Akeley went on his first trip to Africa to what was then called British Somaliland to collect specimens.[41] He returned to Africa in 1905–6 to collect East African animal specimens, including those of elephants.[42] In 1909, Akeley left the Field Museum to begin work at the American Museum of Natural History in New York, where he stayed for the rest of his career. He died in 1926 during his fifth expedition to Africa. In 1920, he published a memoir, titled *In Brightest Africa*, partly to make the case for establishing an extraordinary hall of African habitat dioramas at the New York museum, a hall he hoped would be dedicated to the memory of Theodore Roosevelt but that was eventually named the Akeley Hall of African Mammals, a centerpiece of the museum to this day.

Like Schillings' works, *In Brightest Africa* was written in the tradition of the hunting memoir. Akeley relates his adventures shooting and collecting the dangerous and not-so-dangerous animals he sought for his taxidermic works. Akeley offers plenty of classic accounts of extreme danger, but, again like Schillings, he also steps back at moments to consider changes occurring in Africa, to reflect on the destruction of the game and the encroachment of Western civilization on what he and so many observers saw as the timelessness of Africa. Early in the memoir, for example, he writes about an occasion when he was hunting elephants in the Budongo Forest in Uganda that echoes sentiments expressed by both Schillings and Roosevelt. Akeley describes herds of elephants moving into the forest in the early morning light as the sounds of monkeys echoed through the trees: "I had gone back a million years; the birds were calling back and forth, the monkeys were calling to one another, a troop of chimpanzees in the open screamed, and their shouts were answered from another group inside the forest." As elephants moved in groups and singly, there was, he writes, "one continuous roar of noise, all the wild life joining, but above it all were the crashing of trees and the squealing of the elephants as they moved into the forest on a front at least a mile wide. It was the biggest show I ever saw in Africa."[43] After watching the herd for some time, Akeley finally "caught sight of a fine tusk"—a bull he felt would be a perfect specimen for the New York museum. He could not get a clear shot at the head, so he decided to shoot for the elephant's heart.[44] He fired both barrels; the herd, including the bull, took off. The bull did not get far, though, before he fell. Akeley then saw something he had heard about before but had never witnessed: "My old bull was down on the ground on his side. Around him were ten or twelve other elephants trying desperately with their trunks and tusks to get him on his feet again. They were doing their best to rescue their wounded comrade."[45] Eventually the elephants moved on, leaving the dead bull. Akeley was disappointed to discover that the animal's tusks, though large, were mismatched; the right tusk was 110 pounds but the left only 95 pounds. The base of the left tusk revealed an injury that had apparently happened early in the elephant's life, an injury that had resulted in the tusk growing more slowly and developing a "knotty rib along the entire length."[46] The difference in the tusks and the knotty rib on one made it clear to Akeley that the specimen would not be satisfactory for the display he had imagined. He decided on the spot not to bother preparing the animal in any way and just removed the tusks and sold them for $500.

Despite the disappointing outcome, the story remained important to Akeley because of the behavior of the elephants trying to raise up the fallen bull. The act, which he felt was characteristic of what he called the comradeship shared by elephants, echoed a story he had heard many years earlier from a Major Harrison of two elephants trying to raise a third that had been shot. Realizing the dramatic potential of that image, Akeley created a small bronze in 1913, *The Wounded Comrade*, of two elephants supporting a wounded elephant between them that he hoped might convince potential supporters of the American Museum of Natural History of the artistic potential of full-size taxidermy. When Akeley finally began to consider the composition for a group of taxidermized elephants for the New York museum years later, he stayed with the idea of comradeship. Rather than further developing the idea of the wounded comrade, though, he settled on a family huddled close together, sensing danger. The work features a large and powerful male with his trunk straight out—a posture that foreshadows Iain Douglas-Hamilton's description of Clytemnestra when she became aware of a dead elephant—another male guarding the rear, and a female and her calf in the middle (fig. 6.5). Akeley called the work *The Alarm*.[47] For the piece, Akeley was keen, for the sake of publicity, to involve Roosevelt and met up with the former president and his son while both safaris were under way. Returning to a spot where the Roosevelts had recently observed and photographed a herd, the party found the elephants again and a barrage of shooting began, leaving three adult females

Figure 6.5. Carl Akeley, *The Alarm*, ca. 1914–17. Image 310463, American Museum of Natural History Library.

dead. A small bull calf that stayed standing beside its dead mother was then dispatched by Kermit with a smaller bore rifle.[48]

The Alarm, a depiction of a classic nuclear family, is not exactly what one might expect of an "elephant family" in the wild, but it captures well a particular moment in the history of our ideas about the animals. The elephants here, killed because they were "perfect specimens," are presented sensing but standing up to danger, sensing the hunter, sensing perhaps their own destruction. The work is a monument not simply to the elephant but to its extermination at the hands of man. In his *In Brightest Africa*—a title that attempts to describe Akeley's vision of Africa, which his contemporaries more typically imagined as the "darkest" place—Akeley laments that "it may not be many years before such museum exhibits are the only remaining records of my jungle friends." Using the same historical reference as Schillings, he concludes that "as civilization advances in Africa, the extinction of the elephant is being accomplished slowly but quite as surely as that of the American buffalo two generations ago."[49] Again echoing Schillings, Akeley makes clear that his lasting disappointment with *The Alarm* was that he simply could not find an elephant with bigger tusks. As he notes in a 1912 account of collecting the animals, "The best bull at present in our collection for the group is a young adult standing 11 feet 3 inches at the shoulders with tusks of 100 and 102 pounds respectively." "The world's permanent record of elephant life," he contended, "should contain a specimen that illustrates the fullest development of the African species, the finest living representative of this race of animals. Such an elephant can be secured now, but it will soon be too late, for the remaining monster specimens will be killed for their ivory."[50] The fact that he was also hunting elephants precisely for their ivory does not seem to have registered for Akeley.

Sic transit elephantus

In a 1902 article titled "Doomed" for *Forest and Stream: A Weekly Journal of Rod and Gun*, the occasional contributor T. J. Chapman observes: "Wherever the invading foot of the white man passes, he seems to leave behind him a trail of blood and ruin. In his footsteps later on may spring up amelioration and improvement; but before his face the native races of men and animals give way and disappear. It has been eminently so in this country, and the same process of extirpation has begun in Africa."[51] Chapman quotes what he considers a typical example of sport elephant hunting in Africa from an account of big-game shooting from aboard a boat on the Kassai River, a tributary of the Congo, reported in Frank Vincent's 1895 *Actual Africa*. Vincent explains that one morning, as the boat was rounding a bend in the river his party saw, just in front of the boat, a large elephant "quietly walking across the river and half out of the water." The group grabbed their guns and within a minute had "fired five shots at him, three taking effect, and bringing him to his knees." The elephant got back to his feet, "but soon fell again, and kicked and rolled frantically." It managed to stand again and walk "a few paces further," before it fell again, and "lost by drowning what little of life was left."[52] The account records eleven more elephants being shot at over the course of several days. Only four died close enough to the boat to be recovered. Drawing on ideas we have seen over and over again, Chapman states: "It is, of course, inevitable; it is written in large hand in the book of fate. Ivory is in demand; beside, there is not room for big game, especially of the elephantine order, in a country peopled by civilized men." Noting that just as in North America forests had to give way to farms, that the "red man has been compelled to give place to the white," and that cattle have replaced buffalo, Chapman brings his article to a close arguing that the "same thing is decreed for Africa. The native races must crowd back and finally disappear, and the great beasts of the African forests are doomed. Sic transit elephantus."[53] Arthur Neumann reaches a similar conclusion in the preface to his book: "By all means let elephants and other wild animals be preserved as far as possible. But as, unfortunately, their continued existence is incompatible with the advance of civilization, the only way to do so successfully is by making reserves in places where effective control can be exercised alike over natives and Europeans."[54] We might save some elephants in reserves, Neumann argues, but as for the rest, we might just as well collect their ivory while we can.

At the end of his entry on elephants in the second edition of his *Thierleben* from the late 1870s, Brehm concludes that their "future is clear: they are to be crossed off from the lists of the living."[55] By the beginning of the twentieth century, Brehm's prediction regarding the outcome of the ongoing slaughter of elephants had been realized to such an extent that Schillings' second memoir and Akeley's works feel often more like elegies to past lives than accounts of the present. These are works about the end. Indeed, they often convey the kind of anxiety about the future we are more familiar with today in discussions of extinction and climate change, the rise of populist politics, and the collapse of a consensus on the value of conservation and the importance of repairing the environment. Their works were written in a moment when the ivory trade and sport hunting were in crisis, a time when elephants were used as part of the machinery of empires and seen as only one of the most spectacular markers of imperial success and excess—the animal most suitable for civilized children to ride at one of the new zoos of the urban elite. For Akeley, Schillings, and others, the end of the story of the elephant was at most decades away, and ever since them, we have been watching the disappearance of elephants. *Sic transit elephantus*; thus passes the elephant.

Trails of History

Where this animal has been, there leads a wide path.

—Bertolt Brecht, "Mr. K.'s Favorite Animal"

Beginning in the 1920s during the Weimar Republic through his flight from Germany in 1933 to Denmark, Sweden, Finland, and then the United States, and through his being blacklisted in Hollywood after the war and his final years in East Berlin, the playwright and theatre director Bertolt Brecht was writing small aphoristic stories featuring a man named Herr Keuner, a name that can be translated as Mr. Nobody. Some of the pieces were published separately over the years, but eighty-five were gathered together as a book in 1956, the year of Brecht's death, as *Die Geschichten von Herrn Keuner* (published in English in 2001 as *The Stories of Mr. Keuner*). One of the stories concerns Keuner's favorite animal, the elephant. Keuner describes the creature by listing qualities that echo centuries of thought about elephants in a way that is also quite modern in its humor. According to Keuner, the elephant combines its strength with an intelligence that allows it to accomplish great things. It is good natured and can be both a great friend and a great enemy. While it is both large and heavy, it is also quick. Its trunk leads it to the smallest foods, and it can move its ears and thereby listen only to what it wants. It can grow very old. It is gregarious, and it is both loved and feared. Its skin is so thick, knives break in it, and yet its disposition is gentle. It can become sad and angry. It dies in the thickets, and it loves children and other small animals. It is gray and difficult to see despite its size. It is inedible. It drinks enthusiastically and thus becomes jolly, and it contributes to art by supplying ivory. Beyond these, Mr. Keuner observes, "Wo dieses Tier war, führt eine breite Spur" ("Where this animal has been, there leads a wide path").[1]

For European and American explorers, naturalists, missionaries, and hunters in the eighteenth and nineteenth centuries, traveling through Africa and Asia was always difficult. The problems of hauling baggage and transporting or securing food and water were complicated by weather, diseases, and often only conjectural information about the region into which the party was moving. On maps the country usually lacked detail. As Jonathan Swift observes both amusingly and accurately in his 1733 *On Poetry: A Rhapsody*:

So Geographers in *Afric*-Maps
With Savage-Pictures fill their Gaps;
And o'er unhabitable Downs
Place Elephants for want of Towns.[2]

Swift points to a practice going back to the middle ages of drawing elephants on maps of Africa to indicate the unknown.[3] Elephants marked the unexplored, the unsurveyed, the mysterious; places of both exceeding risk and riches. For the many explorers over the couple of centuries since Swift, both the risk and riches could come from the bodies of elephants. The animals' tusks, meat, skins, and tails all had value, and just following a trail made by a herd of elephants also sped up travel through tall grasses and forests, and even over mountain passes. Carl Akeley notes that "in the forests there are elephant paths everywhere. In fact, if it were not for the elephant paths travel in the forest would be almost impossible, and above the forests in the bamboo country this is equally true. One travels practically all the time on their trails."[4] Some of these trails, he felt certain, had been "used for centuries." Following a herd on the Kinangop Plateau, for example, he used a trail "a little wider than an elephant's foot and worn six inches deep in the solid rock." "It must have taken hundreds of years for the shuffling of elephants to wear that rock away," he explains.[5] In short, following the path of an elephant, or as many elephants as possible, could spell the success of an expedition.

Keuner is talking about more than literal paths, however. For me, the trails left behind include an email I received from a retired elephant trainer for circuses describing a parade of elephants gracefully moving into a long mount, each elephant resting her front feet carefully on the elephant in front of her. Other trails include the face of an elephant in a museum that I had never known about before and the ivory-handled cutlery passed down to me through generations that I have wondered what to do with. I was following footsteps when I watched Brittany, brush in trunk, create a painting for my son. I was on a path when I helped bathe Packy and on another when I spent a lovely afternoon with a zoo historian in a café in London looking at his photo albums of elephants. I was on a trail when I touched Ziggy's skull while Helen and I were photographing elephant bones at the Field Museum in Chicago, just as I was when I went hiking in the Uintah mountains searching for and finding the small flowering plants known as elephant head (*Pedicularis groenlandica*). I was following a path when I visited an elephant sanctuary and met people there who, like people I have met at zoos and circuses, have dedicated themselves to caring for the animals. There is a trail on the refrigerator in my kitchen: a *New Yorker* cartoon about elephant memory clipped out and sent to me by my sister years ago. I was on a trail when I smiled as dear little Lily mouthed my hand. I have found trails in natural histories written thousands of years ago, in medieval bestiaries, in encyclopedias, in imaginative literature, games, and internet memes. They are as present in demands by animal advocates as they are in the range of emotions from boredom to fear to wonder expressed by children feeding bits of food to elephants over a railing at a zoo.[6] I have often stumbled across the trails unexpectedly. While visiting acquaintances once, I found a gigantic plasma television mounted on a taxidermized elephant foot—skin, nails, and a foot pad that seemed so distant from the killed animal from which they had been cut decades earlier. This book is an effort to find and walk on elephant trails and it, too, is a trail.

The frontispiece to the first edition of *In Brightest Africa* is a photograph by Akeley with the caption: "On a Typical Elephant Trail in the Forest" (fig. 7.1). In some ways it might seem a surprising choice for a book of hunting adventures because it is not a portrait of the hunter, not a trophy shot of a dead elephant, lion, or gorilla, but rather a fairly difficult-to-comprehend photograph of an elephant trail in the forest that doesn't really look much like a trail. Barely discernable in the distance are three human figures—Akeley's guides—whom Akeley presents as almost part of the forest itself. In his memoir, Akeley seems to be constantly following physical trails of elephants in the forest, but he also seems to be following abstract ones that tell a deeper, historical story. When he writes of "the shuffling of elephants" wearing down six inches of solid rock over the course of centuries, he is pointing to one of the historical dimensions of elephants' lives—the passing down of knowledge over generations. History is not something that humans alone experience. Elephants living in herds in an Indian or African forest or in a zoo or circus are social creatures who can live long lives. They share experiences and ways of doing things that can then be passed on to others, including other generations.

Catching Apples, Looking at Mirrors, and Knowing Elephants

Elephants do not remember everything. Despite the millennia of wisdom about elephants to which we are heirs, elephants truly do forget things all the time—just like we do. I was talking to an elephant keeper at a zoo once who told a poignant story about elephant memory. Some years earlier, the zoo had an elephant who was quite old and not well. A difficult decision had been

Figure 7.1. Frontispiece to the first edition of Carl Akeley's *In Brightest Africa* (Garden City, NY: Garden City Publishing, 1920).

reached to put the elephant down, and the keeper sent a message out to people who had cared for her over the years, urging them to drop by and spend time with her. Among those who came was a former keeper who picked up an apple and threw it in the open mouth of the elephant, who was waiting expectantly with her trunk raised. The former keeper mentioned that he was glad he dropped by but that he was certain that the elephant did not remember him, to which the keeper who had invited him simply replied that it was difficult to know what an elephant is thinking. When we talked years later, though, he told me that they had stopped throwing food to the elephants over a decade previously and the fact that the elephant raised her trunk and opened her mouth the moment the former keeper showed up made it pretty clear to him that she knew exactly who he was. She just failed to demonstrate her memory in a way that the visitor had hoped or expected. The story is important, I think, because it highlights how our thoughts about elephants—what we think they are like, what we think they are thinking—are always structuring our beliefs about elephants and even what we consider to be our knowledge of them.

One of the most popular arguments made about elephants in recent years, for example—a claim that has appeared in everything from animal rights petitions and natural history magazines and tv programming to zoo signage and even safari websites—is that they are one of the few animals that can recognize themselves in a mirror. It started with a paper published in 1970 by Gordon G. Gallup Jr. with the title "Chimpanzees: Self-Recognition." The article had a two-sentence abstract: "After prolonged exposure to their reflected images in mirrors, chimpanzees marked with red dye showed evidence of being able to recognize their own reflections. Monkeys did not appear to have this capacity."[7] Gallup put two, young, wild-born chimpanzees in cages in an empty room for eight hours a day for two days. He then introduced a full-length mirror. Over the course of eighty hours over eight days, it appeared that the chimps began to regard the mirror differently. Initially, it seemed that the young animals thought their reflections were other animals, but over time they seemed to use the mirror for self-directed activities like cleaning their teeth, blowing bubbles, making faces, or examining parts of their body that they could not otherwise see without the mirror. Gallup then anesthetized the animals and marked them with a red spot that the animals could only see by looking in the mirror. After the animals were awake, they were presented with the mirror again. Gallup found that the "number of mark-directed responses went up dramatically upon re-exposure to the mirror, as did viewing time"; in other words, the chimpanzees seemed to notice the red spots of dye on their bodies when they looked in the mirror.[8] As a control, Gallup performed the final mark test with two other chimps with similar backgrounds who did not have a period of getting used to the mirror, and they did not use the mirror to examine the marks. He also repeated the whole experiment with eight macaques of different species and ages. The monkeys all appeared to continue to regard the image in the mirror as another animal. Gallup concluded that what became known as the mirror self-recognition (MSR) test had shown a "decisive difference" between monkeys and chimps and that the test had provided "the first experimental demonstration of a self-concept in a subhuman form."[9]

After Gallup published the results of his initial experiment, more scientists began to wonder whether other animals might also "pass" the test and might also, therefore, be included in the new and highly exclusive cognitive taxonomy of the self-knowing creatures. Initially, several other kinds of primates passed, suggesting that the close evolutionary or phylogenetic history of primates (including humans) once again separated off this group from all other animals. But when studies were published arguing that dolphins, orcas, elephants, and then European magpies could all recognize themselves in mirrors, it became clear that the kind of self-awareness apparently tested by the MSR was not limited to primates nor even to mammals. The floodgates opened, and now all kinds of animals, from octopuses and cichlids to pigeons, dogs, and ants, have been run through the MSR test as well as through new variants of the experiment designed for animals for whom visual cues are not as important as other senses.

I put the word "pass" in quotation marks at the beginning of the last paragraph because despite how these tests are often presented, the results from almost all the experiments often come with substantial qualifications. In the original experiment, for example, Gallup used preadolescent chimpanzees. It turns out that neither younger nor older chimps do as well with the MSR and that, at best, chimpanzees appear to "pass" the test only about 75 percent of the time. When it comes to elephants, the results are even more murky. To date, there are a total of two scientific papers that have been published. In the first, from 1989, Daniel Povinelli reports of the result of running the Gallup test with two wild-born Asian elephants at the National Zoo in Washington, D.C., by placing a mirror on the outside of the bars of two adjoining stalls for elephants. The subjects were two female wild-born Asian elephants, Shanthi, who was twelve years old at the time of the research, and Ambika, who was thirty-nine.[10] After two weeks of "nearly continuous access to mirrors," Povinelli found, "neither elephant gave any indication of self-recognition."[11] Although the elephants were able, like some monkeys, to use the geometries of the mirrors to access food they could not otherwise see—mirror-guided reaching—they did not display "self-directed responses in which they appeared to be using the mirror to gain otherwise inaccessible information about themselves."[12] They also failed the formal test of self-recognition of exhibiting mark-directed activity in the marker tests.

In the second experiment, the results of which were published in 2006 in the *Proceedings of the National Academy of Sciences* with the title "Self-Recognition in an Asian Elephant," three Asian elephants at the Bronx Zoo in New York, Happy, Maxine, and Patty, were confronted with an eight-foot square mirror mounted to a wall in an off-exhibit, outdoor enclosure. The elephants were housed in pairs. Maxine and Patty were together, and Happy was housed with a young Asian female elephant, Sammy, who was not run through the protocol for reasons that are not specified in the published study.[13] The elephants were given extensive time to familiarize themselves with the mirror. According to the researchers, all three showed a discernible interest in investigating the mirror, and Patty and Maxine both tried to reach over the wall, behind it, and under it with their trunks. None of the elephants exhibited behaviors that observers considered consistent with social interaction with the mirror. Over time, though, and in contrast to Shanthi and Ambika, all three apparently exhibited self-directed behaviors, such as bringing food and eating in front of the mirror. When it came to the all-important mark test, one of the three elephants, Happy, passed the test on the first of the three days of testing. Neither Maxine nor Patty showed any interest in the mark on any of the days, and Happy did not show any interest in the mark on days two and three.[14] Despite the mixed results, the researchers, Joshua Plotnik, Frans de Waal, and Diana Reiss, conclude that the "mark-touching by one elephant is compelling evidence that this species has the capacity to recognize itself in a mirror."[15] Pointing to "strong parallels among apes, dolphins, and elephants in both the progression of behavioral stages and actual responses to a mirror," the authors insist that the experiment "provides compelling evidence for convergent cognitive evolution."[16]

A subtlety of the title "Self-Recognition in an Asian Elephant," is the use of the indefinite article "an." Plotnik, de Waal, and Reiss's paper reports that only one of three Asian elephants seemed to pass the MSR test. The press release from Emory University, where de Waal was a member of the faculty and Plotnik was a researcher, did not hold back, however. According to the release, elephants had "joined a small, elite group of species—including humans, great apes and dolphins—that have the ability to recognize themselves in the mirror." The announcement claims that the "newly found presence of mirror self-recognition in elephants" was thought to be related to "empathetic tendencies and the ability to distinguish oneself from others, a characteristic that evolved independently in several branches of animals, including primates such as humans." In the release, de Waal makes the leap to elephants in general. "As a result of this study," he states, "the elephant now joins a cognitive elite among animals commensurate with its well-known complex social life and high level of intelligence."[17] The story was picked up in the media, and ever since the happy result, the claim that elephants should be regarded differently from other animals because they are self-aware has been made repeatedly. Still, as Gordon Gallup said in a response to the study at the time, "Replication is the cornerstone of science"; alas, more than a decade later, there have not been any studies replicating the Bronx Zoo results, so a more general statement about elephants still seems premature.[18]

In nineteenth-century museum displays, twentieth-century natural history books, Carl Akeley's *The Alarm*, and even the twenty-first-century video game *Zoo Tycoon*, the family life of the elephant is not imagined as a multigenerational, matriarchal group but as a nuclear family—one adult male (usually figured as the leader), one adult female, and their offspring. This way of imagining the family life of elephants makes clear, again, that the ways we think about elephants have as much to do with who we are and what we value as with the lives of actual elephants. Elephants are clearly intelligent, but crows can do as well or better in some kinds of cognitive tests, and the results of the many studies on elephants are perhaps ambiguous because, among other reasons, it can be difficult to distinguish what might be unique to an individual from what is typical for the kind. Why is it, then, that we have imagined elephants as being almost uniquely wise over the many centuries before anyone developed any kind of test to examine that intelligence? I have often thought that oak trees are like elephants in many ways: they are massive, they live long lives but they also can often seem a lot older than they are, they have deep wrinkles, and they have been seen over centuries and centuries as being particularly wise and as having a certain gravitas. I am not saying that oaks and elephants are not wise or serious; I am simply asking why we have thought they might be for so long and what that thought might have to do with human cultures.

So much of what we have thought that we have known about elephants over the centuries—that they are afraid of mice and fight with dragons, that they forget neither a kindness nor an injury, that they mourn their dead—has quite clearly little to do with the actual lives of elephants. Should we not accept, then, that what we think we know about them now will seem ill informed in fifty, a hundred, or five hundred years? Much of the research on elephants in recent decades, for example, has focused on cognition and communication, but at this point so much of the work can only be described as fascinatingly suggestive. There are, of course, large obstacles in studying these phenomena with creatures that cannot exactly be brought into the lab for replicable experiments. In addition, the models that are available to study cognition and communication have typically been developed using species like various primates that are just very different from elephants. From a simple evolutionary standpoint, it is worth reminding ourselves that primates (including humans) are closer cousins to rats, bats, and whales than they are to elephants, and we are clearly only beginning to scratch the surface of understanding the likely very different cognitive world of elephants despite the extensive work that has been done over recent decades.[19]

What will future researchers think of the claims that are being made now about elephant communication? While the idea that elephants keep track of each other over distances of kilometers through infrasound communication is clearly important, what will become of larger claims that suggest networks of communications and interactions spanning great distances and elephants sensing a wide variety of nonbiological sounds (like thunderstorms) from hundreds of kilometers away? Ideas like these, of course, are also part of science, part of learning. Thoughtful researchers pose questions like "can an elephant 'hear' a thunderstorm from very far away?" and then develop experiments to test the questions. But, as we've seen with elephants so often, speculation about elephant minds, emotions, memory, abilities, and the like tend to be way ahead of what we can know, and the public (and often researchers) have often grabbed on to certain ideas about elephants as if they were established facts.

I think back on an event in 2006 when a caregiver at an elephant sanctuary in the United States was killed by an elephant. In the wake of the tragedy, the sanctuary (backed, it was claimed, by scientists and a psychologist advocating for a trans-species approach to understanding animal behavior and psychology) explained the attack as the result of the elephant suffering from post-traumatic stress disorder caused by decades of abuse at the zoo she had left years earlier. The PTSD diagnosis played well in a post-911-media world. Ironically, despite the complexity of the actual psychiatric disorder, the claim that the elephant was suffering from PTSD seemed to provide a simple explanation for the elephant's actions when there were other known and presumably unknown factors that could have also played roles in the event. I'm not saying that the elephant hadn't experienced trauma in her life, nor am I arguing that past trauma could not have played a role in her behavior that day. I want to point out only that the way that the sanctuary explained the event tells us more about how we imagine elephants than about elephants themselves, more about how we inevitably draw on our current ideas about human beings to explain animal behaviors.

Richard Byrne and Lucy Bates point out that elephants appear to have exceptional cognitive mapping skills that help them navigate in very large landscapes, that they can discern quantities well, that they show indications that they can understand what other elephants are thinking, and that they may have larger working memories than humans. At the same time, they make clear that the work in this area is only at its beginning and that "limitations in the data" are evident wherever one turns.[20] These conclusions clearly add to our understanding of the animals. Scientific research moves forward through testing what we believe might be true (hypotheses). In science, ideas always precede knowledge. Current scientific studies about elephants, then, test (disprove or add credence to) our beliefs about the animals—they cannot do more than that. They do not explain why we feel about the animals in the ways we do, why we ask the questions we do when we see elephants at a circus, in a wildlife documentary, or on safari. Once we appreciate the extent to which our thoughts about elephants are always based in our culture, however, it becomes clear that the scientific work focused on elephants over the last fifty years is just the latest chapter in a fascination that has persisted for millennia. But to say that our knowledge is always incomplete and always based in our culture is not the same as saying that we really do not know anything about elephants or that our current ideas about the animals are really no better than one finds in Buffon's or Pliny the Elder's accounts. It is true that our knowledge of elephants is (and will always be) incomplete, that we will always be blind to the full life of these creatures, but we actually do know a great deal more about them today than we ever have.

Near the beginning of this book, I said I wanted to explore the meaning of two words on a label on a crate in a museum in London: "No history." For the museum, the words simply meant that critical pieces of information about the provenance of the specimen were missing. More than anything else, I hope that you will have reached the conclusion through this book that "no history" is an impossibility. Elephants have their own histories and, beyond those, our histories, our identities, our wishes and dreams fill our understanding of what we think about them. The one thing those bones in the museum definitely have is history. This means that when we look at elephants today or in the past, when we try to understand their lives—and what they mean to ours—we have to be aware that we are looking from positions anchored in specific times, places, and cultures. Scientists sometimes struggle with this idea. The worlds of hypotheses, experiments, and theories, though, are also worlds of history and culture. The fact that so many of the questions being asked by scientists today are related to thoughts about the animals that are ancient in origin does not mean that the science is bad or that the questions are unimportant. It means that for millennia elephants have captured the human imagination in ways few other animals have. The artist Charles R. Knight, who sculpted the two African elephant heads for the elephant house at the New York Zoological Park in 1908, also painted a mural of Cro-Magnon man for the American Museum of Natural History in 1920 (fig. 7.2) in which he depicts a group of men painting images of mammoths on cave walls. For Knight and Henry Fairfield Osborn, then president of the American Museum, it made sense to

Figure 7.2. Charles Knight, mural painting of Cro-Magnon cave artists, Font-de-Gaume Grotto, Les Eyzies-de-Tayac. Image 5375, American Museum of Natural History Library.

highlight the artistic accomplishments of Cro-Magnon culture, and it made sense, too, that rather than portraying the cave dwellers painting bison, horses, or rhinos, the prehistoric artists should be shown painting mammoths.[21] For Knight, whose own paintings of mammoths can be seen elsewhere at the American Museum and at the Field Museum in Chicago, contemplating the mammoth and its extinction was a deeply profound artistic experience, an experience he felt connected him to prehistoric artists.[22]

We have been thinking about elephants in certain general ways—that the animals are wise, thoughtful, judicious, and have deep spiritual lives—for a very long time. Still, the elephants of Pliny and Aelian are not the same as the elephants to be found in a medieval bestiary or in the works of Georg Christoph Petri von Hartenfels, William Cornwallis Harris, Alfred Brehm, Theodore Roosevelt, Elwin Sanborn, or Iain and Oria Douglas-Hamilton. Each of these depictions, I think, reveals something about the elephant, but they tell us just as much about different human cultures. When Cynthia Moss wrote about elephant graveyards, she was writing from a very different perspective from that of Percy Powell-Cotton, and yet the idea that elephants retreat to a secret and quiet valley when they are approaching death made sense to both of them and also to their readers. We inevitably bring our cultural backgrounds with us when we interact with the world, but when we think about elephants, about their lives in the past, today, and in the future, I think we must try to be aware of how our ideas about elephants might interfere with our ability to see them for who they are. By attending to the elephant trails all around us, we can begin to tease apart what we know about elephants from what we think we know or just hope.

Because we too often fail to examine how our ideas about our own lives, histories, "nature," and animals can confound one another, we repeatedly end up accepting as credible all kinds of ideas that even on the surface seem to make little actual sense. In recent years, for example, an idea has been circulating that human-elephant conflict is a historically novel phenomenon, caused by a breakdown in elephant culture. Noting a range of "abnormal" behaviors, including elephants "raping and killing rhinoceroses," Charles Siebert pushed this case in a cover story for the *New York Times Magazine* in 2006. Siebert based his article primarily on a claim that there was a consensus among scientists that "where for centuries humans and elephants lived in relatively peaceful coexistence, there is now hostility and violence."[23] I do not know who the scientists were who constituted this consensus, but claiming a behavior is "abnormal" is a notoriously fraught endeavor. In this case, the data for "normality" is stunningly thin. Although we have been imagining the lives of elephants for thousands of years, the field research that constitutes the scientific basis of our understanding of elephants has only been going on for a little over a half century (the lifespan of a single elephant) and has been based on small populations. And, just as in humans, the identification of "pathological" behaviors can often have political overtones—behaviors that do not match our expectations just end up being called "abnormal." In this case, we do know that even a superficial examination of the history of human-elephant interactions points to constant violence between the species. Western thought is marked by the belief in a time when humans lived beside animals, including elephants, in peace; it is not only the Bible that tells this tale. The history of elephant-human interaction, in contrast, has been characterized by consistent avoidance and violence. Even when elephants might appear to be living peacefully with humans as working animals, the training has been either based in "breaking" and dominating the animal through deprivation and physical force or through the repeated use of violence, restraint, and coercion. In his important 2003 work *The Living Elephants: Evolutionary Ecology, Behaviour, and Conservation*, Raman Sukumar confirms that the history of elephants and humans has always been one of intense conflict, a conflict that is even apparent in the history of the worship of Ganesha—a god with both malevolent and benevolent aspects.[24] In the end, we tend to supplicate ourselves before and propitiate gods because we are awestruck by and fear them—we pray to them and make offerings because they might otherwise kill us. To put this another way, our ancestors did not paint mammoths on prehistoric cave walls simply because they found large proboscideans to be naturally sweet, kind, and a large source of meat. I am glad that as I worked on this project, I had the opportunity to meet elephants who helped me appreciate both the terrifying and glorious facets of worship.

This book does not answer the question of why we are so interested in elephants. Carl Jung seems to have had some thoughts on this matter, but I don't know if we will ever really know. Perhaps the answer has something to do with their size, grayness, wrinkles, or faces. Our minds might be hard wired in a way that makes us pay attention to large, potentially dangerous, and mysterious creatures like elephants. By becoming more aware of the elephant trails around us, though, I hope we can begin to be more careful in our thinking about these animals. Statements beginning "Elephants are . . ." have started many arguments we have about elephants. In that ellipsis are contained thoughts that can only be understood as coming from the speaker's particular and limited—blind—view of the world. We would do more good, I believe, if we were more modest in our claims about these animals. Our history of thinking about these remarkable entities should make us cautious in what we claim to know about them and what is best for them.

Tears

In a sense, the elephants whose trails this book follows are all "lasts." The nascent awareness in the decades surrounding the end of the nineteenth century of the vulnerability of species to total extinction marks the beginning of a broader comprehension of both geologic time and the recent planetary impact of human activity. Writers, scientists, and even general audiences are increasingly resigned; the question for many has become not whether elephants and other creatures and plants of the planet will go extinct but when. I, too, have often found it difficult to avoid a tone of lament as I have been writing this book—it seemed that everywhere I turned all I found was death and suffering. A different kind of historian, one who works in much deeper time, might make the reasonable point that many, many species of proboscidean that once walked this planet are gone. Along with the mammoth and the mastodon, the traces of the Phosphatherium, Deinotherium, Platybelodon, Anancus, and literally hundreds of other kinds of elephant-like animals are only to be found today in fossils.[25] In the context of Earth's full history, the species of African and Asian elephants alive today are newcomers, first showing up as recognizable members of their species in the last one-tenth of one percent of Earth's history. With that said, the five million years modern elephants have been around is still fourteen times longer than *Homo sapiens* have been thinking about them, admiring them, and killing them.

Probably everyone who has seen more than a handful of nature documentaries in recent decades has gotten used to the often-strained efforts of the documentarians to try to end their narratives on an optimistic note. Perhaps room for that kind of optimism is possible at the end of this project; after all, we have been predicting the demise of elephants for well over a century and they are still here. Indeed, in some areas, elephants are clearly exceeding the carrying capacity of the constricted lands they still have available, and in others their numbers inevitably lead to heightened human-elephant conflict. I am sorry to say that the fact that elephants are not gone yet, despite the warnings of decades, provides little reason to be sanguine about their future. With larger and highly mobile animals like elephants, extinction can be a slow process. While a flightless bird living on an island—like a dodo—might disappear in a matter of years under the assault of changing climate, humans, their domestic animals, and other co-colonizers like rats, woolly rhinos and mammoths survived centuries of dramatic changes and human hunters before succumbing forever. Still, if there is a will to keep pockets of "wild"—if also monitored, fenced, and otherwise controlled—elephants alive in the world, sufficient resources will probably be marshaled to protect a few thousand elephants into the foreseeable future. While almost all of the species that go extinct every day are not noticed by humanity, elephants, as both profound ideas and beings that live long lives in a wide range of environments, have advantages in this world and are likely to be around for some time to come.

In 2001, around the time when I began my research for this project, I found myself after hours at the zoo near where I live. I was there with a curator at the zoo to see an elephant whose companion for close to forty years had recently died. The being on the other side of the bars from me looked utterly devastated, sunken into herself. She had not slept since the other elephant had died over a week earlier; she hadn't eaten much either. We tried to understand what might be going through her mind—certainly a deep sense of loss, of missing, and probably also utter confusion. It was winter and during that season both of the elephants spent almost all of their time in the small elephant house in close proximity. This elephant had experienced death before, long ago when she was very small and lived in Africa. The elephant standing before me, into whose eyes I then looked, had experienced something again that changed her world, and the twenty-four-hour-a-day watch by those who cared for her seemed insufficient help. A new, young elephant arrived before long and slowly the older animal returned to her life. She, too, though, would die five years later at the age of forty-six.

That day the curator and I talked about grieving.[26] We had both read accounts of the significance of death to elephants, about elephant mourning, about so-called elephant death rituals, about the emotional lives of elephants. We had read Iain and Oria Douglas-Hamilton and Cynthia Moss. We knew, too, some of the history of elephants described as crying or weeping. Jeffrey Moussaieff Masson and Susan McCarthy's *New York Times* bestseller *When Elephants Weep: The Emotional Lives of Animals* was published in the mid-1990s and had made a big impression on many people. Through anecdotes, through attention to historical accounts of animals, through empathy, and through common-sense arguments, the book makes a case that the emotional lives of animals are richer, deeper, and more nuanced than most readers imagined and than most scientists would assert. In some ways, Masson and McCarthy make the emotional lives of animals simply more comprehensible by arguing that there are good reasons to believe that many of the emotional states we experience—and that we believe are common among humans—are also present among nonhuman animals. Perhaps not every species of animal experiences every emotion we know, but a wide range of emotional states might reasonably be inferred for all kinds of creatures. At the same time, Masson and McCarthy argue that we should not assume that the emotions we experience as humans are an exhaustive catalog of all possible emotions. They urge us to recognize that while a chimpanzee's emotional life, for example, might be similar to ours, it might also be quite different. *When Elephants Weep* asks simply that readers embrace the idea that the nonhuman animals of the world have complex emotional lives.

Masson and McCarthy draw on often ancient accounts of animal emotion, but their starting point seems in many ways to have been Charles Darwin's 1872 *The Expression of the Emotions in Man and Animals*.[27] In that work, Darwin has, in fact, very little to say about elephants, but he does observe that "the Indian elephant is known sometimes to weep." Darwin had two sources. He points first to a keeper of the elephants at the Zoological Gardens of London who had claimed to see "tears rolling down the face" of an older female Asian elephant when she was separated from a younger companion.[28] Darwin reports that he then went to the zoo to confirm the claim, but that he did not see either the Asian or African elephants weeping. Darwin's second source was James Tennent's account of the elephant corral in which a lone male elephant lay on the ground "uttering choking cries, with tears trickling down his cheeks."[29] In this case, too, Darwin sought confirmation and wrote to correspondents in Ceylon who had also witnessed catching corrals, but no one reported ever seeing elephants crying or in tears. In the end, though, Darwin accepted the claims of the keeper and Tennent because he found them plausible and they at least confirmed each other.

Darwin and Tennent, however, are not the only sources for the idea that elephants weep. Masson and McCarthy point to both Gordon-Cumming's account of taking experimental shots at a wounded elephant while hunting in South Africa and an account from Slim Lewis about an elephant crying after being beaten as part of training. Beyond these, many readers, even today, might remember that in the opening pages of Jean de Brunhoff's 1931 *Babar the Elephant*, Babar is illustrated crying while standing on top of his dead mother, who has just been shot by a hunter.[30] But stories of elephants weeping go back to classical authors. Although Pliny reserves grieving and crying for mankind, writing "On man alone of living creatures is bestowed grief," Aelian describes the tears of elephants forced to leave their homelands.[31] He claims that although elephants can be taken from their native lands and tamed they never really forget their homes and also asserts that while many die of grief, others have even gone blind because of the "floods of tears" they had shed.[32] In the seventeenth century, Edward Topsell picked up Aelian's story, stating that elephants have such a "wonderful love to their own Countrey" that memories of it cause them to "send forth tears."[33] Buffon, however, was more reserved about tears, noting only that both ancient and more modern authors have claimed that when elephants come on the corpses of fellow elephants they "bedew them with tears."[34]

Masson and McCarthy note pithily that "tears are not grief, but tokens of grief."[35] They assemble a wide variety of accounts in order to make their point that regardless of whether elephants are physically able to produce emotional tears—and there is a scientific literature that suggests they cannot—elephants can undoubtedly be unhappy and, in a certain way, weep.[36] As a historian, I see the writings of Aelian, Topsell, Buffon, Gordon-Cumming, Tennent, Darwin, de Brunhoff, and even Masson and McCarthy as both telling us about elephants and telling us about our own interests in the meaning and importance of tears and weeping. Do I believe elephants grieve? I know they do; I saw it while standing before that grieving elephant at my local zoo. With that said, I recognize that my thoughts about grief are also part of who I am and very much a part of the times in which I live. I am, like Brehm, Roosevelt, or Barnes, seeing only a part of the truth of elephants. In a discussion of animal emotions, the ethologist Marc Bekoff observes that "even if joy and grief in dogs are not the same as joy and grief in chimpanzees, elephants, or humans, this does not mean that there is no such thing as dog joy, dog grief, chimpanzee joy, or elephant grief." Bekoff insists that it is simply not enough in these days to say circumspectly that "elephants *seem* to feel grief"; we should just acknowledge that they do.[37] When it comes to tears, though, I think that whether or not elephants weep emotional tears, the fact that so many people think they do (and have for a very, very long time) should challenge us to think about how our thoughts about elephants are always also thoughts about ourselves.

~

In a story some fifteen hundred years older than Antoine Galland's translation of the adventures of Sindbad, Aelian describes a sacred spot at the foot of Mount Atlas that hints at tales of elephant graveyards. There, amid deep and dense forests and beautiful pastures, elephants gathered in their old age. Below arching trees, where a spring of the purest drinking water always flowed abundantly, the elephants lived out the remaining years of their lives in peace, unassaulted by local peoples and cared for by "certain gods of the district who are lords of wood and valley." Then one day, a distant king who coveted the old elephants' huge tusks sent three hundred hunters to slaughter the sacred herd. As the party neared where the elephants lived, however, "a pestilence seized them and laid them low." As happens in these kinds of stories, one hunter survived to tell the tale, and what his listeners all realize from hearing it is that "elephants were beloved of the gods."[38] At the end of this book about elephants, this story from Claudius Aelianus, a Roman born around 170 CE in the town of Praeneste, comforts me. Despite all the evidence, I like the thought that elephants are "beloved of the gods." One other line from Aelian's account comforts me, too: "Were I to pass over a piece of cleverness on the part of an Elephant, someone will say that I failed through ignorance to record it."[39] I know there is so much about elephants and our thoughts about them that I have not recorded in this book, but I will offer just one last, and again quite old, story that I think holds a piece of wisdom for our times.

In 202 BCE in what is now Tunisia in North Africa, one of the most famous battles in world history was fought between Rome and Carthage. The Roman army of twenty-nine thousand infantrymen with a cavalry of six thousand was under the leadership of Publius Cornelius Scipio. The Carthaginian army was led by Hannibal and had a larger infantry of thirty-six thousand, a smaller cavalry of four thousand, and more than eighty war elephants. The stakes were high. As the Greek historian Polybius puts it, "To the Carthaginians it was a struggle for their own lives and the sovereignty of Libya; to the Romans for universal dominion and supremacy."[40] At the end of what became known as the Battle of Zama, tens of thousands of the soldiers lay dead, but Scipio's force was victorious. Critical to his success, as recorded in Polybius, was his response to the advance of Hannibal's elephants. As the elephants charged, Scipio's army sounded horns and trumpets to frighten the animals, and the foot soldiers repositioned themselves to open up straight alleys through their lines in front of the charging elephants. Despite the efforts of the mahouts, some of the confused and frightened elephants turned back against the Carthaginian infantry, but more simply continued down the open alleys between the Roman infantrymen. Many were attacked from the sides by Romans throwing spears, but others simply moved as quickly as they could to safety beyond the battle lines of the great armies.

Millennia after Zama, military historians continue to debate the lessons of the historic battle. My interests are different from theirs, and the message of this event for me is not Scipio's brilliance in overcoming a larger infantry and a phalanx of elephants. What I see, in the end, is that the elephants sought only to escape the conflicts of humanity as they moved toward the open alleys leading beyond Scipio's lines. As I think about the elephants of today, besieged on all sides with shrinking and changing habitats, by the ongoing ivory trade, and by continued exploitation, I hope we can create and they can find new paths to safety.

Introduction

1. The story is included in the Udana, one of the scriptural works of the Pāli Canon.

2. John Godfrey Saxe, "The Blind Men and the Elephant," *The Poems* (Boston: Osgood, 1872), 259.

3. Saxe, "The Blind Men and the Elephant," 261.

4. The story, of course, has been used to other ends. It has been argued that although the story might make us skeptical about our supposed knowledge of transcendent truths or gods, it can also be taken to mean that even if people struggle to gain deeper wisdom about life and the universe, that deeper truth is present (the elephant "exists"). Others insist that one can accept the story and still argue that a particular religion is true; one simply has to believe that the whole truth has been revealed in the word of a prophet or god and is not based on the limited and fallible perceptions of a human, that the beliefs come from a complete revelation of the whole "elephant."

5. Helen Keller, "My Animal Friends," *Zoological Society Bulletin* 26, no. 5 (1923): 114. My thanks to Madeleine Thompson of the Wildlife Conservation Society for her support of this project and for pointing me to this article.

6. Romain Gary, *The Roots of Heaven*, trans. Jonathan Griffin (1958; rpt., New York: Time and Simon and Schuster, 1964), 6. My thanks to Noëlle Pujol and Violette Pouillard for their encouragement that I read this book.

7. Gary, *The Roots of Heaven*, xv.

8. Hans Hermann Schomburgk, *Wild und Wilde im Herzen Afrikas; Zwölf Jahre Jagd- und Forschungsreisen* (Berlin: Fleischel, 1910), 278, my translation. For more on Schomburgk, see Bernhard Gissibl, *The Nature of German Imperialism: Conservation and the Politics of Wildlife in Colonial East Africa* (New York: Berghahn, 2016).

9. For more on Hagenbeck's Tierpark, see Nigel Rothfels, *Savages and Beasts: The Birth of the Modern Zoo* (Baltimore, MD: Johns Hopkins University Press, 2002).

10. Schomburgk called the captured elephant "Jumbo," recalling the much more famous African elephant who lived in the Paris and London zoos from the 1860s to 1880s before he was sold to P. T. Barnum to tour the United States. In Rome, the elephant became known as Toto.

11. See Spartaco Gippoliti, "One Elephant, a Museum Specimen and Two Colonialisms: The History of M'Toto from German Tanganyika to Rome," *Museologia scientifica*

8 (2014): 67–70. See also Noëlle Pujol, *Jumbo/Toto, histoires d'un éléphant*, 67 minutes, DCP (Pickpocket Production and Noëlle Pujol, 2016). My thanks to Spartaco and Noëlle for their collaborative engagement on the history of Jumbo / Toto.

12. It is difficult to know Toto's age when he was captured. In the photograph by Schomburgk, he looks to me to be about three to four years old, so I am using 1905 as his year of birth.

13. I had doubts about the authenticity of the photograph, wondering whether it was a composition of two images (one of a young elephant and one of a dead elephant). When I contacted a descendent of Schomburgk's who owned the original glass plate of the image, she confirmed that the image appeared to have been taken as it was in the field.

Chapter 1 • First among Monsters

1. I use the language of the original British Museum Act of 1753, even though the act was repealed in 1963 when the Natural History Museum officially separated from the British Museum, because the specimens I consider here are from before 1963 and because the intent of preserving the collection for future generations remains a hallmark of the institution.

2. There are many versions and translations of *Arabian Nights* and the Sindbad stories. The Lang edition does not have the same goals as Richard Burton's classic 1885 *The Book of the Thousand Nights and a Night*, but it benefits from a more pleasant literary style. In the case of the events surrounding the elephant graveyard, Lang stays reasonably close to Galland's account. For comparison, see the anonymous and fairly literal eighteenth-century English translation of Galland's work titled *Arabian Nights Entertainments: Consisting of One Thousand and One Stories . . . Translated from the Arabian Manuscript into French by M. Galland*, 3 vols. (London: C. Cooke, 1706).

3. Edmund Heller, "Nature's Most Amazing Mammal: Elephants, Unique Among Animals, Have Many Human Qualities When Wild That Make Them Foremost Citizens of Zoo and Circus," *National Geographic Magazine* 65 (June 1934): 752.

4. Lawrence G. Green, "Seeking the 'Ivory Valley': African Expedition Formed to Hunt for the Legendary Place Where Elephants Die," *New York Times*, January 8, 1933. Heller's and Green's articles were published in the wake of the very successful films *Tarzan the Ape Man* (1932) and *Tarzan and His Mate* (1934), both of which centrally deployed the idea of the elephant graveyard.

5. Trader Horn, *Trader Horn; Being the Life and Works of Aloysius Horn, an "Old Visitor"* (New York: Literary Guild of America, 1927), 115–16.

6. Percy Horace Gordon Powell-Cotton, *In Unknown Africa: A Narrative of Twenty Months Travel and Sport in Unknown Lands and among New Tribes* (London: Hurst and Blackett, 1904), 379.

7. Homer, *The Odyssey*, trans. A. T. Murray (Cambridge, MA, Harvard University Press, 1919), 269.

8. Vergil, *The Aeneid*, trans. Allen Mandelbaum (Berkeley: University of California Press, 1971), 162.

9. Perhaps Vergil intended for readers to doubt the truth of Aeneas's account or perhaps we are to question what Aeneas learned about his future and the future of Rome. One argument claims that the gates are only for shades and that because Aeneas and the Sibyl are real and not shades—because they are, then, "false shades"—they must pass

through the gate of ivory. See Nicholas Reed, "The Gates of Sleep in *Aeneid* 6," *Classical Quarterly*, n.s., 23, no. 2 (1973): 311–15. A different approach points to Vergil's commitments to Platonism and insists that what comes through the gate of ivory only seems false to most in the world but that it is possible to discern the truth behind what is seen. As T. J. Haarhoff frames this argument, "What issues from [the gate of horn] are *umbrae*, shadows, true in the sense that they form the reality of ordinary men. Opposed to them are the *insomnia*, visions, that form the real but hidden truth, but are fantastic to all except the Seer who penetrates the outward husk of things" ("The Gates of Sleep," *Greece & Rome* 17, no. 50 [1948]: 90).

10. Ernest Leslie Highbarger, *The Gates of Dreams: An Archaeological Examination of Vergil, "Aeneid" VI, 893–899* (Baltimore, MD: Johns Hopkins University Press, 1940), 28. See also Haarhoff, "The Gates of Sleep," and W. F. J. Knight, "A Prehistoric Ritual Pattern in the Sixth Aeneid," *Transactions and Proceedings of the American Philological Association* 66 (1935): 256–73. In the notes to his translation of *The Odyssey*, Arthur Murray suggests that these materials were chosen because of likenesses in ancient Greek between the words for "horn" and "fulfill" (κέρας and κραίνω) and "ivory" and "deceive" (ἐλέφας and ἐλεφαίρομαι) (269). Perhaps Homer was simply making puns, but it also seems that the gates conceit predates Homer.

11. See Howard Hayes Scullard, *The Elephant in the Greek and Roman World* (London: Thames and Hudson, 1974), 32–37.

12. Pliny, *The Natural History of Pliny*, vol. 2, trans. John Bostock and H. T. Riley (London: Henry Bohn, 1855), 245.

13. Pliny, *The Natural History of Pliny*, 244.

14. Pliny, *The Natural History of Pliny*, 244.

15. Pliny, *The Natural History of Pliny*, 247.

16. Aelian, *On the Characteristics of Animals*, vol. 1, trans. A. F. Scholfield (Cambridge, MA: Harvard University Press, 1958), 347–49. The thoughts of Pliny, Aelian, and other classical writers about elephants would echo in bestiaries, natural histories, and other works all the way into nineteenth century. In his letter calling for the fortification of crumbling sculptures of elephants along the Via Sacra in Rome—which he thought particularly regrettable given that by comparison "these animals live in the flesh more than a thousand years"—Cassiodorus (c. 485–c. 585 CE), for example, argues that the elephant exceeds all other animals in intelligence and that this is demonstrated by the "adoration which it renders to Him whom it understands to be the Almighty Ruler of all" (*The Letters of Cassiodorus: Being a Condensed Translation of the "Variae Epistolae" of Magnus Aurelius Cassidorus Senator*, trans. Thomas Hodgkin [London: Henry Frowde, 1886], 442).

17. Iain Douglas-Hamilton and Oria Douglas-Hamilton, *Among the Elephants* (New York: Viking, 1975), 293.

18. Douglas-Hamilton and Douglas-Hamilton, *Among the Elephants*, 293–94.

19. Douglas-Hamilton and Douglas-Hamilton, *Among the Elephants*, 296.

20. Douglas-Hamilton and Douglas-Hamilton, *Among the Elephants*, 295.

21. Douglas-Hamilton and Douglas-Hamilton, *Among the Elephants*, 300.

22. Douglas-Hamilton and Douglas-Hamilton, *Among the Elephants*, 300.

23. Cynthia Moss, *Elephant Memories: Thirteen Years in the Life of an Elephant Family* (1988; rpt., Chicago: University of Chicago Press, 2000), 270.

24. Moss, *Elephant Memories*, 270.

25. Moss, *Elephant Memories*, 270–71.

26. See Karen McComb, Lucy Baker, and Cynthia Moss, "African Elephants Show High Levels of Interest in the Skulls and Ivory of Their Own Species," *Biology Letters* 2, no. 1 (2006): 26–28.

27. Joyce H. Poole and Peter Granli, "Signals, Gestures, and Behavior of African Elephants," in *The Amboseli Elephants: A Long-Term Perspective on a Long-Lived Mammal*, ed. Cynthia J. Moss, Harvey Croze, and Phyllis C. Lee (Chicago: University of Chicago Press, 2011), 109–24.

28. Moss, *Elephant Memories*, 269–70.

29. Robert Steele, *Medieval Lore from Bartholomew Anglicus* (London: Alexander Moring, 1905), 153.

30. Geoff Boxshall notes that if there were an average of just one copepod per cubic liter of ocean water, the number of copepods on the planet would be over 1,347,000,000, 000,000,000,000 (i.e., 1.35×10^{21}) individuals ("Preface to the Themed Discussion on 'Mating Biology of Copepod Crustaceans,'" *Philosophical Transactions of the Royal Society of London, series B: Biological Sciences* 353 no. 1369 [1998]: 669).

31. I am indebted to Rudi for his comradeship over decades. Marks of his intellectual and personal generosity can be found throughout this book.

32. My thanks for this translation to my sister, Janet Rothfels, who has been a wonderful partner through so much of this project. For the Latin, see D. R. Shackleton Bailey, ed., *Anthologia latina* (Stuttgart: Teubner, 1982), 128–29.

33. Edward Topsell, *The History of Four-Footed Beasts, Serpents, and Insects* (London: Cotes, 1658), 756.

34. Ernest W. Shaw, "Where Do Dead Elephants Go?" *New York Times*, October 20, 1929.

35. See, for example, "Josephine, Elephant Pygmy, Dies," *Indiana (PA) Evening Gazette*, March 13, 1943.

36. I base this interpretation on the gestural database for African elephants developed by field researcher Joyce Poole and her associates. The database documents the gestures of wild savannah elephants and may not be an ideal resource to use with an elephant caught in the Congo as an infant.

Chapter 2 • Afraid of Mice

1. For an insightful reading of taxidermy, see Rachel Poliquin, *The Breathless Zoo: Taxidermy and the Cultures of Longing* (State College: Penn State University Press, 2012).

2. Emmanuel Levinas argues that "the skin of the face is that which stays most naked, most destitute. It is the most naked, though with a decent nudity. It is the most destitute also: there is an essential poverty in the face; the proof of this is that one tries to mask this poverty by putting on poses, by taking on a countenance. The face is exposed, menaced, as if inviting us to an act of violence. At the same time, the face is what forbids us to kill" (*Totality and Infinity: An Essay on Exteriority*, trans. Alphonso Lingis [The Hague: Nijhoff, 1969], 85–86). Levinas was not talking about animals—for him, animals have a face only through analogy, only insofar as they remind us of the human—but it was clear to me that day in the museum that I was seeing the face of an elephant. See also Jacques Derrida, *The Animal That Therefore I Am*, ed. Marie-Louise

Mallet and trans. David Wills (New York: Fordham University Press, 2008), and Matthew Calarco, "Faced with Animals," in *Radicalizing Levinas*, ed. Peter Atterton and Matthew Calarco (Albany: State University of New York Press, 2010), 113–33.

3. Pliny, *The Natural History of Pliny*, vol. 3, trans. John Bostock and H. T. Riley (London: Henry Bohn, 1855), 52.

4. The Elephant Sanctuary has since taken down the memorial pages dedicated to Barbara.

5. *A True and Perfect Description of the Strange and Wonderful Elephant Sent from the East-Indies* (London: F. Conniers, 1675).

6. The main scholarly sources that referenced elephants were the *Collectanea rerum memorabilium*, a compressed third-century version of Pliny's *Natural History*; the *Hexameron*, St. Ambrose's fourth-century account of the six days of creation; the *Etymologiae*, Isidore of Seville's seventh-century encyclopedia; and the *Physiologus*, a text that dates back to the second or third centuries that demonstrates how the natural histories of mammals, birds, insects, fish, plants, and even rocks reveal stories and lessons from the Bible.

7. Willene B. Clark, *A Medieval Book of Beasts: The Second-Family Bestiary* (Woodbridge, UK: Boydell, 2006), 7.

8. James Thomson, *The Seasons* (London: Longmans, 1852), 108–9.

9. For a full account of the elephant in the bestiaries, see George Druce's classic "The Elephant in Medieval Legend and Art," *Journal of the Royal Archaeological Institute* 76, no. 1 (1919): 1–73. In his study, Druce provides a slightly augmented translation of the entry on elephants from what is known as the Harley Manuscript 3244 housed in the British Library.

10. Among the more famous elephants in Europe in these years were the early sixteenth-century Hanno, given to Pope Leo X and sketched sympathetically by Raphael, and the mid-seventeenth-century Hansken who toured widely around Europe and was sketched by Rembrandt. For a fascinating account of Hanno, see Silvio Bedini, *The Pope's Elephant: An Elephant's Journey from Deep in India to the Heart of Rome* (New York: Penguin, 1997).

11. Edward Topsell, *The Historie of Foure-Footed Beastes* (London: Iaggard, 1607), 196. See also Conrad Gessner, *Thierbuch: Das ist ein kurtze Beschreybung aller vierfüssigen Thieren* (Zurich: Froschouwer, 1583), 75r.

12. According to Topsell, "They are most chast, and keepe true vnto their males without all inconstant loue or seperation [and] take their veneriall complements, for the continuation of their kind, and neuer aboue thrice in all their daies" and are "modest and shamefast in this action, for they seeke the Desarts, woodes, and secret places for procreation," "turne their heads towards the east, but whether in remembrance of Paradise, or for the Mandragoras, or for any other cause, I cannot tell," and "goe into the Water to the belly and there calve for feare of the Dragon" (*The Historie of Foure-Footed Beastes*, 197–98).

13. "They haue also a kinde of Religion, for they worshippe, reuerence, and obserue the course of the Sunne, Moone, and Starres" (Topsell, *The Historie of Foure-Footed Beastes*, 207).

14. Topsell, *The Historie of Foure-Footed Beastes*, 193. Gessner similarly notes that "the head for the elephant is very large. There are smaller ears and eyes than one would

expect, given the proportions of such great size. Varro compares their eyes to pigs' eyes" (*Historiae animalium*, vol. 1 [Frankfurt: Cambieriano, 1602], 379, my translation).

15. Thomas Jefferson, *Notes on the State of Virginia* (London: Stockdale, 1787), 71.

16. See Louise E. Robbins, *Elephant Slaves and Pampered Parrots: Exotic Animals in Eighteenth-Century Paris* (Baltimore, MD: Johns Hopkins University Press, 2002).

17. Georges-Louis Leclerc de Buffon, *Natural History: General and Particular by the Count de Buffon*, trans. William Smellie, 2nd ed., vol. 6 (London: Strahan and Cadell, 1785), 4–6.

18. Buffon, *Natural History*, 6–7.

19. Buffon, *Natural History*, 10–11.

20. Buffon, *Natural History*, 15.

21. Buffon, *Natural History*, 17.

22. Buffon, *Natural History*, 25–26.

23. Buffon applauds the "exquisite discernment" of the Marquis de Montmirail, who collected and translated materials for Buffon from Italian and German sources (*Natural History*, 72).

24. Buffon's conclusion, for example, that elephants mated face-to-face was based on accounts equating the size of the elephant's penis to that of a horse; from that inaccurate piece of information, Buffon concluded that mating could only be successful with the female on her back.

25. Buffon, *Natural History*, 71.

26. Buffon, *Natural History*, 48.

27. Buffon's most important source here appears to have been von Hartenfels: "The elephant moves his eyes in a dignified manner. . . . Elephant eyes resemble human eyes, not only in their shape, but also in their expressiveness. For truly they are humanlike; serious and temperate, they show wisdom, justice, moderation, and other qualities of the human mind. Truly, so great is the seriousness in their eyes, that by their gaze alone it is clear that the elephant is the king of the animals, and what is more, that the elephant approaches human sensibilities. The elephant can distinguish among those unfamiliar who approach him, from those who are trifling, arrogant, and impudent and those who are moderate and serious, whom he contemplates with greater pleasure" (*Elephantographia curiosa* [Leipzig, 1723], 29, my translation).

28. For brief biographical accounts of many of the elephants that came to Europe in these centuries, see Stephan Oettermann, *Die Schaulust am Elefanten: Eine Elephantographia Curiosa* (Frankfurt: Syndicat, 1982).

29. In *The Animal Kingdom Arranged According to Its Organization*, vol. 1 (London: G. Henderson, 1834), Cuvier writes: "After studying them for a long time, we have not found [their intelligence] to surpass that of the Dog, or of many other carnivorous animals. Naturally of a mild disposition, Elephants live in herds, which are conducted by old males. Their food is strictly vegetable" (132).

30. See Lynn K. Nyhart, *Modern Nature: The Rise of the Biological Perspective in Germany* (Chicago: University of Chicago Press, 2009).

31. Brehm wrote the first edition while serving as the first director of the Zoological Garden of Hamburg, a position that undoubtedly gave him a good sense for the kinds of information the public wanted. While preparing the second edition he was serving as the first director of the new Berlin Aquarium located on the main boulevard through

the city, Unter den Linden. Beginning with the third edition (1890–93) and as part of broader orthographic changes in the German Empire in the 1890s, the spelling for animal, "Thier," was changed to "Tier," with subsequent editions then bearing the title, *Brehms Tierleben*. For more on Brehm, see Nyhart, *Modern Nature*.

32. John Corse, "Observations on the Manners, Habits, and Natural History, of the Elephant," *Philosophical Transactions of the Royal Society of London* 89, no. 2 (1799): 31–55. The paper was presented by the president of the society, Sir Joseph Banks, on January 24, 1799. On May 23, 1799, a second paper was presented with the title "Observations on the Different Species of Asiatic Elephants, and Their Mode of Dentition"; see *Philosophical Transactions of the Royal Society of London* 89, no. 2 (1799): 205–36.

33. In "Observations on the Manners, Habits, and Natural History, of the Elephant," Corse writes that "the elephant has been declared to possess the sentiment of modesty in a high degree; and, by some, his sagacity was supposed to excite feelings for the loss of liberty, so acute, as to cause him to refuse to propagate his species while in slavery, lest he should entail on his progeny a fate similar to his own; whilst others have asserted, that he lost the power of procreation in the domestic state" (31–32).

34. Corse, "Observations on the Manners, Habits, and Natural History, of the Elephant," 53.

35. Corse, "Observations on the Manners, Habits, and Natural History, of the Elephant," 55.

36. See Alfred Edmund Brehm, *Illustrirtes Thierleben*, vol. 2 (Hildburghausen: Bibliographisches Institut, 1865), 688.

37. For an insightful account of James Emerson Tennent's political career, see Jonathan Jeffrey Wright, "'The Belfast Chameleon': Ulster, Ceylon and the Imperial Life of Sir James Emerson Tennent," *Britain and the World* 6, no. 2 (2013): 192–219. After publishing *Ceylon: An Account of the Island Physical, Historical, and Topographical with Notices of Its Natural History, Antiquities and Productions*, 2 vols. (London: Longman, 1859), Tennent published a volume limited to just the natural history of the island titled *Sketches of the Natural History of Ceylon with Narratives and Anecdotes Illustrative of the Habits and Instincts of the Mammalia, Birds, Reptiles, Fishes, Insects, &c., including a Monograph of the Elephant and a Description of the Modes of Capturing and Training It* (London: Longman, 1861)—essentially a reprinting of the natural history sections of *Ceylon*. Finally, in 1867, his sections on elephants in *Ceylon* were published alone as *The Wild Elephant and the Method of Capturing and Taming It in Ceylon* (London: Longman, 1867).

38. See Wright, "'The Belfast Chameleon,'" 199, 214.

39. Brehm, *Illustrirtes Thierleben*, 694.

40. Brehm relates an anecdote of Tennent's that points to the limits of Tennent as an honest reporter. According to Tennent, one evening when he was out riding near Kandy, his horse became unsteady. Coming toward them was a tame elephant, working alone without a mahout, struggling with a heavy log in its trunk. Realizing that there was not enough room for both him and the horse and rider, the elephant, according to Tennent, immediately dropped the log and shuffled itself backward into the forest beside the path to make way for the skittish, and for Tennent, less admirable horse: "On seeing us halt, the elephant raised his head, reconnoitered us for a moment, then flung

down the timber and forced himself backwards among the brushwood so as to leave a passage, of which he expected us to avail ourselves" (*Ceylon*, 2:283). The story does not make a great deal of sense—domestic elephants do not "work alone" carrying logs. That Tennent relates the story as if he had directly observed the scene thus makes clear that he was not a reliable reporter. That Brehm relied on him for his own work, though, suggests that Brehm found Tennent convincing.

41. Tennent, *Ceylon*, 2:279.

42. Because of the larger format of Brehm's book, Tennent's forty-three pages (*Ceylon*, 2:335–78) are reduced to ten pages in the *Illustrirtes Thierleben* (699–709).

43. Tennent, *Ceylon*, 2:335.

44. Tennent, *Ceylon*, 2:347.

45. Tennent, *Ceylon*, 2:348.

46. Tennent, *Ceylon*, 2:349.

47. Tennent, *Ceylon*, 2:353.

48. Tennent, *Ceylon*, 2:357. When the older elephant eventually died, Tennent reports that he arranged for its skeleton to be sent to the Belfast Natural History and Philosophical Society Museum.

49. Tennent, *Ceylon*, 2:358

50. Tennent, *Ceylon*, 2:365.

51. Tennent, *Ceylon*, 2:363–64.

52. Tennent, *Ceylon*, 2:376.

53. Georg Schweinfurth, *Im Herzen von Afrika: Reisen und Entdeckungen im Centralen Aequatorial-Afrika während der Jahre 1868–1871*, 2 vols. (Leipzig: Brockhaus, 1874), 1:476.

54. Schweinfurth, *Im Herzen von Afrika*, 2:293–94, my translation.

55. Alfred Edmund Brehm, *Brehms Thierleben: Allgemeine Kunde des Thierreichs*, 2nd ed., vol. 3 (Leipzig: Bibliographisches Instituts, 1877), 501, my translation.

Chapter 3 • A Serpent for a Hand

1. *Tabellarische Übersichten des Hamburgischen Handels*, Zusammengestellt von dem handelsstatistischen Bureau (Hamburg: Kümpel, 1850–1913).

2. Between 1850 and 1913, the total value of imports coming into Hamburg from the sea increased from the equivalent of 253,648,900 Reichsmarks in 1850 to 4,716,186,110 Reichsmarks in 1913 (an over eighteen-fold increase), reflecting a growth in the total weight of shipments from 556,730,900 kilograms in 1850 to 16,548,410,300 kilograms in 1913 (an almost twenty-nine-fold increase). For ease and clarity, in preparing the figures presented here, I have converted weight measurements from Centner and Doppelzentner to kilograms and pounds and currencies from Hamburger Banco Marks and Vereinsthalern to Reichsmarks.

3. The amount of one product imported did decrease over the period. In 1850, 335,550 kilograms of baleen or whalebone (the plates of filtering material in the mouths of some whales which were used principally in the clothing industry in the nineteenth century) were brought into the city; by 1913 that number had dropped to 81,100 kilograms. The amount of hippopotamus and walrus teeth imported also trended slightly downward over the period, but the year-over-year records varied dramatically, with extraordinary peaks in the 1870s and 1910.

4. Between 1850 and 1913, 6,164,600 kilograms of seal and sea lion skins were imported. The average weight of skins varied year by year (depending on the kinds of seals harvested) from just over a kilogram to over five kilograms. Based on an average of five years of harvests for which numbers of pelts were counted, I have used 2.85 kilograms as an average pelt weight to arrive at 2,171,694 pelts.

5. The total weight of ivory recorded by the bureau was 24,095,312 pounds.

6. There are reasons to think that closer to 750,000 elephants were killed in service of the Hamburg market during this time. First, the average weight of tusks has been debated for a very long time. In the third edition of *Brehms Tierleben* (Leipzig: Bibliographisches Institut, 1891), Eduard Pechuel-Loesche argues that average weight per tusk may actually have been closer to eight kilograms or 730,000 elephants (37). Beyond that, every account of elephant hunting I have read records plenty of elephants that managed to escape with wounds that appeared fatal to the hunter. In any case, in the twentieth century, and especially in the second half of the century, the average weight of tusks declined substantially. Whereas nineteenth-century hunters tended to reserve their efforts and munitions for large elephants with substantial tusks, ivory hunters have become less discriminating over the last century. A study of confiscated tusks in Dar-es-Salaam in the late 1970s and the Hong Kong and Japanese markets in the 1980s suggests that average tusk weights in the late twentieth century had diminished to between five to ten kilograms. See Ian S. C. Parker and Esmond Bradley Martin, "How Many Elephants Are Killed for the Ivory Trade?" *Oryx* 16, no. 3 (1982): 235–39.

7. See Ian S. C. Parker's *The Ivory Trade* (Washington, DC: Department of Fish and Wildlife, 1979). This project was undertaken at the direction of Iain Douglas-Hamilton. Parker underscores the importance of taking into account the fact that most ivory imported into receiving countries involved tusks already present in the trade; one could not, therefore, just sum imports into the major trading states because one would likely then be double or triple counting the same tusk. More recent estimates of the number of elephants killed in Africa through poaching (the proportion of illegally killed elephants, or PIKE) suggest ten to twenty thousand elephants are being killed every year for their ivory. See, for example, Severin Hauenstein, Mrigesh Kshatriya, Julian Blanc, Carsten F. Dormann, and Colin M. Beale, "African Elephant Poaching Rates Correlate with Local Poverty, National Corruption and Global Ivory Price," *Nature Communications* 10, no.1 (2019): 2242.

8. The most famous of the hunters were figures like Charles John Andersson, Samuel White Baker, Walter Dalrymple Maitland "Karamojo" Bell, Roualeyn Gordon-Cumming, William Cornwallis Harris, Arthur Neumann, Percy Horace Gordon Powell-Cotton, Thomas William Rogers, George P. Sanderson, Carl Georg Schillings, Hans Hermann Schomburgk, Frederick Courteney Selous, Chauncey Hugh Stigand, and James Sutherland.

9. In Joseph Conrad's 1899 novel *Heart of Darkness*, the reason Kurtz, the renowned agent in charge of the station deep in the jungle, was such a success is because he had managed to compel indigenous people to devote themselves to collecting ivory.

10. Georg Schweinfurth, *Im Herzen von Afrika: Reisen und Entdeckungen im Centralen Aequatorial-Afrika während der Jahre 1868–1871*, vol. 2 (Leipzig: Brockhaus, 1874), 27, my translation.

11. Indeed, celluloids, the first forms of plastic, were developed in the 1850s and 1860s partly as a replacement for ivory.

12. Jason Colby in his *Orca: How We Came to Know and Love the Ocean's Greatest Predator* (New York: Oxford University Press, 2018) presents a fascinating case study of whales that parallels the rise of concern for elephants. Colby traces how attitudes about orcas changed radically over the course of just the few decades after the 1960s. At the beginning of the period, they were entirely unprotected by legislation and were regularly shot and harassed for fun and because they were regarded as pests. Just a few decades later, however, the whales had become icons of the Pacific Northwest, protected through legislation and defended by major animal rights groups. Again, as in the case of elephants, many of those who ended up working hardest to protect orcas were formerly people involved in hunting and the capture of the animals.

13. Alfred Edmund Brehm, *Illustrirtes Thierleben: Eine Allgemeine Kunde des Thierreichs*, vol. 2 (Hildburghausen: Bibliographisches Institut, 1865), 697, my translation.

14. Roualeyn Gordon-Cumming, *Five Years of a Hunter's Life in the Far Interior of South Africa, with Notices on the Native Tribes, and Anecdotes of the Chase, of the Lion, Elephant, Hippopotamus, Giraffe, Rhinoceros, &c.*, 2 vols. (New York: Harper Brothers, 1850), 2:14–15.

15. Gordon-Cumming, *Five Years of a Hunter's Life in the Far Interior of South Africa*, 2:15.

16. On the systematic denial of the presence of large numbers of Africans helping settlers and later safari hunters in Africa, see Edward I. Steinhart's *Black Poachers, White Hunters: A Social History of Hunting in Colonial Africa* (Oxford: James Currey, 2006).

17. William Cornwallis Harris, *Wild Sports of Southern Africa* (London: John Murray, 1839), 203.

18. By "salaam-like," Gordon-Cumming no doubt intends to refer to the physical gesture of the trunk touching the head, but his use of the term is stunningly ironic in this context.

19. Gordon-Cumming, *Five Years of a Hunter's Life in the Far Interior of South Africa* 2:15–16.

20. Brehm, *Illustrirtes Thierleben*, 697, my translation.

21. Brehm, *Illustrirtes Thierleben*, 697–98, my translation.

22. "Krank geschossenes Wild," *Die Gartenlaube: Illustriertes Famlienblatt* 46 (1895): 788, my translation.

23. "Krank geschossenes Wild."

24. "Krank geschossenes Wild."

25. Even if Harris's techniques were broadly the same as Gordon-Cumming's—they were using similar weapons, they both would generally initially shoot to lame the animals, and they were wasteful in their almost constant hunting—Harris has rarely received the condemnations that have been directed at Gordon-Cumming. I think this is partly due to the fact that Harris does not linger on the death throes of animals in his descriptions.

26. Gordon-Cumming, *Five Years of a Hunter's Life in the Far Interior of South Africa*, 1: 261.

27. Gordon-Cumming, *Five Years of a Hunter's Life in the Far Interior of South Africa*, 1:264.

28. Gordon-Cumming, *Five Years of a Hunter's Life in the Far Interior of South Africa*, 1:265–66.

29. Gordon-Cumming, *Five Years of a Hunter's Life in the Far Interior of South Africa*, 1:297–299.

30. Gordon-Cumming, *Five Years of a Hunter's Life in the Far Interior of South Africa*, 1:299–301.

31. Gordon-Cumming's obituary in the *New York Times* notes his "ability in telling large stories." See "Death of Gordon Cumming, the African Lion-Hunter," *New York Times*, April 17, 1866.

32. Gordon-Cumming, *Five Years of a Hunter's Life in the Far Interior of South Africa*, 1: 302.

33. Jeffrey Wright shows well why Tennent may have been so critical of other British residents in Ceylon; see "'The Belfast Chameleon': Ulster, Ceylon and the Imperial Life of Sir James Emerson Tennent," *Britain and the World* 6, no. 2 (2013): 192–219.

34. James Emerson Tennent, *Ceylon: An Account of the Island Physical, Historical, and Topographical with Notices of Its Natural History, Antiquities and Productions*, 2 vols. (London: Longman, 1859), 2:323.

35. Tennent, *Ceylon*, 2:323.

36. Tennent, *Ceylon*, 2:326, emphasis in the original.

37. Tennent, *Ceylon*, 2:326; Gordon-Cumming, *Five Years of a Hunter's Life in the Far Interior of South Africa*, 2:4–5.

38. Tennent, *Ceylon*, 2:324.

39. "Elephant Shooting in Ceylon," *Harper's New Monthly Magazine* 8, no. 48 (1854): 758.

40. Samuel White Baker, *The Rifle and the Hound in Ceylon* (London: Longman, 1854), 7–8.

41. Baker, *The Rifle and the Hound in Ceylon*, 8–9.

42. Baker, *The Rifle and the Hound in Ceylon*, 9.

43. Baker, *The Rifle and the Hound in Ceylon*, 108.

44. Baker, *The Rifle and the Hound in Ceylon*, 109–10. I discuss this hunt further in my essay "Mammoths in the Landscape," in *Routledge Handbook of Human-Animal Studies*, ed. Susan McHugh and Garry Marvin (London: Routledge, 2014), 10–22.

45. Tennent, *Ceylon*, 2:326–27.

46. Baker, *The Rifle and the Hound in Ceylon*, 267–68.

47. Baker, *The Rifle and the Hound in Ceylon*, 270–71.

48. Theodore Roosevelt, *African Game Trails: An Account of the African Wanderings of an American Hunter-Naturalist* (London: John Murray, 1910), ix. In act 5, scene 3, of *Henry IV, Part II*, Ancient Pistol proclaims, "A foutre for the world and worldlings base! / I speak of Africa and golden joys." Roosevelt included a copy of Shakespeare in his "Pigskin Library," the fifty-nine volumes he had specially bound and packed for the trip.

49. That the trip was a tribute to Roosevelt is obvious in everything from the many cartoons appearing in newspapers and magazines at the time to the program for the

welcome home dinner celebrating Roosevelt in New York on June 22, 1910. A digitized version of the program is available at https://www.biodiversitylibrary.org/item/88604#page/9/mode/1up. On Roosevelt's long-standing passion for natural history, see Darrin Lunde, *The Naturalist: Theodore Roosevelt, a Lifetime of Exploration, and the Triumph of American Natural History* (New York: Broadway, 2016).

50. Roosevelt, *African Game Trails*, 2–3.

51. Theodore Roosevelt, "Primeval Man; and the Horse, the Lion, and the Elephant," in *A Book-Lover's Holidays in the Open* (New York: Scribner's 1916), 192–93.

52. Roosevelt, "Primeval Man," 193.

53. The original photograph shows Roosevelt standing next to Leslie Tarlton, one of the organizers of the safari. Tarlton does not appear in the frontispiece of *African Game Trails*.

54. Roosevelt, "Primeval Man," 202.

55. Roosevelt, *African Game Trails*, 244.

56. Roosevelt, *African Game Trails*, 245.

57. Roosevelt, *African Game Trails*, 247–48.

58. Roosevelt, *African Game Trails*, 248.

59. Roosevelt, *African Game Trails*, 249.

60. Roosevelt, *African Game Trails*, 249.

61. Roosevelt, *African Game Trails*, 253.

62. In his essay "Books for Holidays in the Open" in *A Book-Lover's Holidays in the Open*, Roosevelt describes Macaulay: "The historian whom it has been the fashion of the intellectuals to patronize or deride showed a much sounder philosophy and an infinitely greater appreciation of and devotion to truth than was shown by the loquacious apostle of the doctrine of reticence" (266).

63. Thomas Babington Macaulay, *The Lays of Ancient* Rome (London: Longmans, 1842), 187.

64. Roosevelt, *African Game Trails*, 241.

65. Arthur H. Neumann, *Elephant Hunting in East Equatorial Africa: Being an Account of Three Years' Ivory-Hunting under Mount Kenia and among the Ndorobo Savages of the Lorogi Mountains, including a Trip to the North End of Lake Rudolph* (London: Rowland Ward, 1898).

66. Neumann, *Elephant Hunting in East Equatorial Africa*, 55.

67. Neumann, *Elephant Hunting in East Equatorial Africa*, 56.

68. Neumann, *Elephant Hunting in East Equatorial Africa*, 57.

69. Neumann, *Elephant Hunting in East Equatorial Africa*, 58.

70. Neumann, *Elephant Hunting in East Equatorial Africa*, 59.

71. Neumann, *Elephant Hunting in East Equatorial Africa*, 59.

72. Neumann, *Elephant Hunting in East Equatorial Africa*, 59–60.

73. Neumann, *Elephant Hunting in East Equatorial Africa*, 60.

74. Neumann, *Elephant Hunting in East Equatorial Africa*, 61.

75. Neumann, *Elephant Hunting in East Equatorial Africa*, 62.

76. Roosevelt, *African Game Trails*, 306.

77. Roosevelt, *African Game Trails*, 62.

78. Roosevelt, *African Game Trails*, 53.

79. See, among many other works, Frederick Courteney Selous, *A Hunter's Wanderings in Africa: Being a Narrative of Nine Years Spent amongst the Game of the Far Interior of South Africa* (London: Macmillan, 1907), C. H. Stigand, *Hunting the Elephant in Africa: And Other Recollections of Thirteen Years' Wanderings*, with an introduction by Theodore Roosevelt (New York Macmillan, 1913), John Henry Patterson, *The Man-Eaters of Tsavo and Other East African Adventures* (London: Macmillan, 1907), Charles John Andersson, *The Lion and the Elephant* (London: Hurst and Blackett, 1873), Percy H. G. Powell-Cotton, *A Sporting Trip through Abyssinia* (London: Rowland Ward, 1902), and George P. Sanderson, *Thirteen Years among the Wild Beasts of India* (London: Allen, 1879).

80. Roosevelt, *African Game Trails*, 3, 6.

81. Midgely's critique of Gordon-Cumming in her argument regarding elephanticide could be extended to Harris, Baker, and Roosevelt, but I think she might have given Neumann a pass for his clear-eyed awareness that he was in Africa only to shoot elephants and collect their ivory and not to test himself in a match that was essentially entirely one sided. (*Animals and Why They Matter* [Athens: University of Georgia Press, 1983], 15).

82. Some of the criticisms directed at Roosevelt are rather difficult to parse and seem to come down to questions of personality as much as anything. Lunde writes, "Interestingly, it was the elite European big-game-hunting fraternity that most loudly condemned Roosevelt's scientific collection" (*The Naturalist*, 253).

83. James Sutherland, *The Adventures of an Elephant Hunter* (London: Macmillan, 1912), 15.

84. I have written about Delia Akeley and her first elephant in "Trophies and Taxidermy," in *Gorgeous Beasts: Animal Bodies in Historical Perspective*, ed. Joan Landes, Paula Young Lee, and Paul Youngquist (State College: Penn State University Press, 2012), 117–36. See also Steinhart, *Black Poachers, White Hunters*, 115.

85. George Orwell, *Shooting an Elephant and Other Essays* (New York: Harcourt, Brace, 1950), 8.

86. Hunters, themselves, also criticized each other. Neumann, who showed that elephant hunting could also just be a practice of slaughter, faced criticism from those he mockingly described as "superior sportsmen" (*Elephant Hunting in East Equatorial Africa*, viii) who disapproved of his willingness to kill as many elephants as he could, male or female.

87. "Long Attacks Roosevelt. Says the Effect of the Big Hunt is to Brutalize Boys," *New York Times*, May 27, 1909. See also Lunde, *The Naturalist*, 173–76, and 229–32.

88. For an enigmatic study of the disappearance of wildlife (the end of the game) and the persistence of a deeply nostalgic view of a settler colonial past that included safari hunting, see Peter H. Beard's troubled *The End of the Game: The Last Word from Paradise* (New York: Doubleday, 1977). Gregg Mitman discusses Beard—and also Iain Douglas-Hamilton and Cynthia Moss—in his thoughtful treatment of animal and human celebrity in "Pachyderm Personalities: The Media of Science, Politics, and Conservation," in *Thinking with Animals: New Perspectives on Anthropomorphism*, ed. Lorraine Daston and Gregg Mitman (New York: Columbia University Press, 2005), 175–95.

Chapter 4 • The Most Friendly Creature

1. See Dick Blau and Nigel Rothfels, *Elephant House* (State College: Penn State University Press, 2015).

2. H. L. Mencken, *Damn! A Book of Calumny* (New York: Goodman, 1918), 84.

3. Mencken, *Damn!*, 80.

4. Shana Alexander, "Belle's Baby—225 Pounds and All Elephant," *Life Magazine*, May 11, 1962, 104–20.

5. Carl Zuckmayer, quoted in Herman Reichenbach, "Carl Hagenbeck's Tierpark and Modern Zoological Gardens," *Journal of the Society for the Bibliography of Natural History* 9, no. 4 (1980): 573; Günter H. W. Niemeyer, *Hagenbeck: Geschichte und Geschichten* (Hamburg: Hans Christians, 1972), 7. For full-length accounts of Hagenbeck, see Nigel Rothfels, *Savages and Beasts: The Birth of the Modern Zoo* (Baltimore, MD: Johns Hopkins University Press, 2002), Lothar Dittrich and Annelore Rieke-Müller, *Carl Hagenbeck (1844–1913): Tierhandel und Schaustellungen im Deutschen Kaiserreich* (Frankfurt: Peter Lang, 1998), and Eric Ames, *Carl Hagenbeck's Empire of Entertainments* (Seattle: University of Washington Press, 2009). My thanks to Herman Reichenbach for his guidance regarding Hagenbeck over many years. See also Herman Reichenbach, "A Tale of Two Zoos: The Hamburg Zoological Garden and Carl Hagenbeck's Tierpark," in *New Worlds, New Animals: From Menagerie to Zoological Park in the Nineteenth Century*, ed. Robert J. Hoage and William A. Deiss (Baltimore, MD: Johns Hopkins University Press, 1996), 51–62.

6. Hornaday began his career as a taxidermist in 1873 for Ward's Natural Science Establishment in Rochester, New York, and became the chief taxidermist of the US National Museum in 1882. In 1889 he was named the founding director of the National Zoo in Washington, DC.

7. On the matter of trees, however, in 1919 the New York Zoological Society did devote an entire issue of its bimonthly *Bulletin* to efforts by the society, led by Hornaday's close associate Madison Grant, to preserve the last stands of Pacific redwoods. See Madison Grant, "Saving the Redwoods: An Account of the Movement during 1919 to Preserve the Redwoods of California," *Zoological Society Bulletin* 22, no. 5 (1919): 90–106.

8. Hornaday explains in the *Twelfth Annual Report of the New York Zoological Society* (New York: Office of the Society, 1908) that "our visitors throw upon our walks and grounds at least twenty times more rubbish and waste paper than any attendance of visitors would dare to throw down in any European park" and that "the worst offenders are the lower class aliens, who insist upon doing here what they never dared to do in their home countries" (84).

The idea that immigrants were responsible for savaging the songbird population came out of anecdotal data Hornaday collected for a report on the disappearance of birds across the United States. He had written to correspondents in every state and territory, and based on the replies he received, he concluded that there were four main causes for the disappearance of birds in the US: the "slaughter of edible birds," the "destruction of birds for millinery purposes," "the scourge of egg-collectors," and "hunting contests" in which teams, or "sides," competed to see how many birds and animals could be killed in a given period. Strikingly, even though the cause of decline most frequently cited by his respondents was sportsmen, Hornaday avoids discussing hunting except to note the particularly egregious case of "side-shoots." Meanwhile, even though only twelve respondents listed "Italians, and others, who devour songbirds" as a cause of decline, Hornaday argues forcefully for the severe punishment of immigrants, "creoles," and others caught killing songbirds to eat them. Hornaday refers in his text to letters claiming that "Italians are beginning to kill the small song birds," that "the Italians are destroying the small singing birds," that "there is a gang of Italians in Providence who gather for food everything that has feathers or fur," and that "Italians are scouring the country, particularly on Sundays" ("The Destruction of Our Birds and Mammals: A Report on the Results of an Inquiry," *Second Annual Report of the New York Zoological Society* [New York: Office of the Society, 1899], 78, 85–86).

9. William T. Hornaday to Carl Hagenbeck, February 20, 1903, William T. Hornaday and W. Reid Blair Outgoing Correspondence 1895–1940 (hereafter OC), Wildlife Conservation Society Archives (hereafter WCSA). For more on the antelope house, see my "The Antelope Collectors," in *Zoo Studies and the New Humanities*, ed. Tracy McDonald and Daniel Vandersommers (Kingston, Ontario: McGill-Queen's University Press, 2019), 45–64.

10. William T. Hornaday to Carl Hagenbeck, March 26, 1903, OC, WCSA.

11. William T. Hornaday to Carl Hagenbeck, April 20, 1903, OC, WCSA.

12. William T. Hornaday to Carl Hagenbeck, October 6, 1903, OC, WCSA.

13. William T. Hornaday to Carl Hagenbeck, February 23, 1904, OC, WCSA.

14. Carl Hagenbeck to William T. Hornaday, March 7, 1904, William T. Hornaday and W. Reid Blair Incoming Correspondence and Subject Files 1895–1940 (hereafter IC), WCSA.

15. Carl Hagenbeck to William T. Hornaday, March 24, 1904, IC, WCSA.

16. Also in shipment were animals bound for the St. Louis zoo, for which Hornaday would arrange transshipment, including "1 female Elephant and Baby; 5 Tortoises, large ones; 4 boxes Snakes; 2–3 boxes Lizards; 1 box 2 Swans; 1 box 3 Baboons; 1 box Sootey Monkeys, Caletriz Monkeys etc., 2 boxes 4 Chimpanzees, 1 box 1 Gibbon" (Carl Hagenbeck to William T. Hornaday, June 9, 1904, IC, WCSA).

17. Carl Hagenbeck to William T. Hornaday, June 20, 1904, IC, WCSA.

18. The elephant Gunda was then assigned to Keeper Frank Gleason, who, with a raise on August 1, 1904, was earning $60 per month. My thanks to Madeleine Thompson of the Wildlife Conservation Society who located correspondence, dated July 29, 1904, from Hornaday to Madison Grant (OC, WCSA) requesting raises for Gleason, four other keepers, and a janitor. The keepers' adjusted salaries ranged from $55 to $65 per month; the janitor got the largest raise, from $40 to $50 per month.

19. William T. Hornaday to Carl Hagenbeck, July 11, 1904, OC, WCSA.

20. Carl Hagenbeck to William T. Hornaday, July 15, 1904, IC, WCSA. In Hornaday's letter of July 11, he writes: "In my conversation last Saturday with Kodah Bux, he cleared up the misunderstanding which had arisen between him and the interpreter in the previous conversation. The interpreter made a mistake in understanding Kodah Bux to say that this animal had killed a man before coming here. He did not say so; but, on the contrary, assured me that this elephant is a good-tempered animal. You will not therefore need to do anything at all about the request contained in my previous letter, and I am sorry about the annoyance that the mistake will have caused you by this time."

21. William T. Hornaday to Heinrich Hagenbeck, August 17, 1904, OC, WCSA.

22. William T. Hornaday, "Our First Elephant," *Zoological Society Bulletin* 15 (October 1904): 182–83.

23. "The elephant is full of vigor, and while quite good-tempered and tractable, he seems to think it is his duty to destroy everything in and about his stall, that can be broken" (William T. Hornaday, *Ninth Annual Report of the New York Zoological Society* [New York: Office of the Society, 1905], 63).

24. "Gunda the Good Elephant," *Washington Post*, October 15, 1905.

25. Ellen Velvin, "'Animals with a History'—Dear Old Gunda the Elephant," *New York Times*, February 25, 1906. Velvin was not a member of the newspaper's staff but an independent writer and a fellow of the Zoological Society of London. Hornaday frequently turned to her for a variety of writing assignments, even recommending her to Hagenbeck as a potential ghost writer for Hagenbeck's memoir.

26. A second boy took part in the photographs that day but is obscured in this picture by the smaller girl.

27. Most taxonomists at the time recognized at least two species of African elephants, what are now recognized as *Loxodonta africana* Blumenbach (1797) and *Loxodonta cyclotis* Matschie (1900). Throughout the twentieth century, there was substantial disagreement that has still not been completely resolved today about whether it makes sense to distinguish African elephants into separate species, partly because of the substantial overlaps in their ranges and the questionable conservation status of these "hybrids."

28. Payne was a member of the American aristocracy of the period. Named for Commodore Oliver Hazard Perry, a relative of his mother's, Payne graduated from Yale University, where he was a member of the elite Skull and Bones secret society. Subsequently, he was connected to the American Tobacco Trust, Standard Oil, and US Steel. Hornaday regularly called on the wealthy members of the New York Zoological Society such as Payne and Payne's brother-in-law, William C. Whitney, to purchase the more expensive animals for the collection. In these cases, he was assiduous to match the grandiosity of the animal to the status of the donor, making sure that the animal was worthy of the donor's name, which was typically added to the enameled sign describing the animal at its cage.

29. Hornaday, *Ninth Annual Report of the New York Zoological Society*, 69–70.

30. *Tenth Annual Report of the New York Zoological Society* (New York: Office of the Society, 1906), 83. The *Zoological Bulletin* put the "total net sum realized" at $1,375" but I think the figures in the annual report are likely more accurate. According to the *Bulletin*, over the course of the summer, "ten thousand tickets for the ponies and carts were sold, and twenty-five hundred for the elephant" ("The Riding Establishment," *Zoological Society Bulletin* 16 [January 1905]: 203).

31. A strong performance in 1906 followed, with $1,503.32 in receipts (*Eleventh Annual Report of the New York Zoological Society* [New York: Office of the Society, 1907], 77).

32. "Elephant Seizes Woman. 'Sweetheart' of Gunda in New York Has Close Call," *Chicago Daily Tribune*, August 12, 1906.

33. In "'Animals with a History'," Velvin describes a fairly benign experiment of this nature in which she tested Gunda to see if he could the difference between a coin and a button of about the same size with his trunk. He could.

34. See "An Embezzling Elephant," *Youth's Companion Magazine*, April 25, 1907, 204.

35. "Elephant Attacks Keeper. Gunda, Trick Animal of the Bronx Zoo, Crushes Hoffman's Ribs," *New York Times*, July 29, 1907.

36. "Gunda to Be Let Off," *New York Times*, July 30, 1907.

37. See "Ardent Elephant Hugs Aged Woman," *Washington Post*, September 15, 1907.

38. "Makes Pet of Big Indian Elephant: Aged Mrs. Hawes Feeds Him Cookies and Says She's Not a Bit Afraid," *New York Times*, September 15, 1907.

39. "Captive Elephant's Love for a Frail Woman," *Washington Post*, October 20, 1907.

40. William T. Hornaday, "The African Elephant," *Zoological Society Bulletin* 19 (October 1905): 237–38. See also "New African Elephants," *Zoological Society Bulletin* 26 (July 1907): 349–50. Hornaday notes that he was looking for donors to pay for the animals: "Incidentally it may be added that they are yet to be paid for, and therein lies a fine opportunity for the making of two grand gifts, each in the sum of $2500, wherewith to pay for these animals. To-day their cost is very reasonable. In a few years they will be the most gigantic and awe-inspiring beasts in Greater New York, and eventually they will be worth at least $8000 each. If no ill fortune should befall Kartoom, he should attain a shoulder height of eleven feet and a weight of 12,000 pounds. Such a gift would do credit to any donor and he will be accredited to the first person who sends $2500 as his purchase money. His mate costs the same amount, and is equally eligible" (350).

41. The zoo seems to have tried to rechristened Alice as Luna, but her old name persisted. See William T. Hornaday, "A Sacred Elephant," *Zoological Society Bulletin* 31 (October 1908): 454–55.

42. "Old Keeper Brings Elephant to Terms: Alice Greets Her Friend with Joy after Spreading Terror in Bronx Park," *New York Times*, September 20, 1908.

43. *Thirteenth Annual Report of the New York Zoological Society* (New York: Office of the Society, 1909), 35.

44. "The Elephant House," *Zoological Society Bulletin* 35 (October 1909): 563.

45. William T. Hornaday, *Popular Official Guide to The New York Zoological Park*, 11th ed. (New York: Zoological Society, 1911), 91, emphasis in the original.

46. "The Elephant House," *Zoological Society Bulletin* 31 (October 1908): 451.

47. "From Jungle to Bronx Palace: Elephants in the Zoo Are Looking Forward to a Grand Moving Day This Month," *New York Times*, November 8, 1908.

48. "The Elephant House," *Zoological Society Bulletin* 35 (October 1909): 563.

49. William T. Hornaday, "The Zoological Park of Our Day," *Zoological Society Bulletin* 35 (October 1909): 544.

50. The *Zoological Society Bulletin* 24 (January 1907) republished a review of the zoo titled "The New York Zoo" by F. G. Aflalo that had appeared in the London *Outlook*. Aflalo concludes: "The Park, which approaches completion, is already a marvelous achievement; and when Mr. Hornaday rests from his labors, the science of the outdoor menagerie, conducted on lines at once popular and humane, will know no higher expression than it will find in the glades and valleys of the Bronx" (326).

51. William T. Hornaday, *Popular Official Guide to The New York Zoological Park*, 11th ed. (New York: New York Zoological Society, 1911), 89.

52. William T. Hornaday, "The Zoological Park of Our Day," *Zoological Society Bulletin* 35 (1909): 543.

53. Hornaday, *Popular Official Guide*, 91.

54. "Gunda Tries to Kill Keeper. Bronx Zoo Elephant Barely Misses Crushing Walter Thuman," *New York Times*, July 29, 1909.

55. "Zoo Elephant Tries to Kill Its Keeper. Knocks Him Down with Its Trunk and then Gores Him in the Leg," *New York Times*, July 13, 1912.

56. "Zoo Elephant Tries to Kill Its Keeper."

57. "Bronx Zoo Elephant Chained for 2 Years: Visitors Stirred to Pity for Gunda, Bound in His Cage after Attacking Keeper," *New York Times*, June 23, 1914.

58. "Gunda's Exercise Is the One-Step. Bronx Keeper Says Big Elephant's 'Weaving' Keeps Him Contented," *New York Times*, June 25, 1914.

59. "Gunda's Case Again Considered," *New York Times*, June 25, 1914.

60. "To Chain the Elephant for Life Is Unpardonable," letter to the editor, *New York Times*, June 26, 1914.

61. "Hornaday on Gunda," letter to the editor, *New York Times*, June 27, 1914.

62. "Times Readers Protest against Gunda's Imprisonment," *New York Times*, July 19, 1914.

63. "It's Now Up to Gunda. Elephant's Chains Will Be Removed as Soon as He Becomes Safe," *New York Times*, August 14, 1914.

64. "Gunda Again Comes into View," *New York Times*, August 15, 1914.

65. "Put Double Chains on Gunda Again. Bronx Zoo Elephant Is Condemned to Stand and 'Weave' All Day in His Pen. Keepers Say He Likes It." *New York Times*, January 12, 1915.

66. "Bullet Ends Gunda, Bronx Zoo Elephant: Dr. Hornaday Ordered Execution Because Gunda Reverted to Murderous Traits," *New York Times*, June 23, 1915. See also Nigel Rothfels, "Trophies and Taxidermy," in *Gorgeous Beasts: Animal Bodies in Historical Perspective*, ed. Joan Landes, Paula Young Lee, and Paul Youngquist (State College: Penn State University Press, 2012), 117–36.

67. "Bullet Ends Gunda, Bronx Zoo Elephant."

68. Raymond L. Ditmars, "Individual Traits of Elephants," *Zoological Society Bulletin* 18, no. 1 (1915): 1187. Ditmars is incorrect about the date of the attack on Thuman; he meant 1912.

69. James Emerson Tennent, *The Wild Elephant and the Method of Capturing and Taming It in Ceylon* (London: Longmans, 1867), 21.

70. *Twentieth Annual Report of the New York Zoological Society* (New York: New York Zoological Society, 1916), 67.

71. "Bullet Ends Gunda, Bronx Zoo Elephant."

72. "Gunda's Obituary," *New York Times*, June 24, 1915.

73. "Bronx Zoo Elephant Chained for 2 Years.".

74. Elwin R. Sanborn, "The Case in Hand," *Zoological Society Bulletin* 53 (September 1912): 910–11.

75. Just how "wild" these animals are remains open to debate; at the very least, they are carefully monitored. I have written more about the complicated meaning of reintroduction in "Re(Introducing) the Przewalski's Horse," in *The Ark and Beyond: The Evolution of Zoo and Aquarium Conservation*, ed. Ben A. Minteer, Jane Maienschein, and James P. Collins (Chicago: University of Chicago Press, 2018), 77–89.

76. There are estimates that over a million racoons are now living in the wild in Germany, descendants of racoons that were brought to Germany in the early twentieth century.

77. *Twentieth Annual Report of the New York Zoological Society* (New York: New York Zoological Society, 1916), 37.

78. The male elephant Ziggy at the Brookfield Zoo in Chicago was chained in a stall, like Gunda, by a foreleg and a hind leg, for almost thirty years from 1941 to 1970.

Chapter 5 • A Descendant of Mastodons

1. The same article with minor variations appeared in many papers, some now long defunct, including the *Tensas Gazette* of Saint Joseph, Louisiana, the *Altoona Tribune* of Altoona, Kansas, the *Bucklin Banner* of Bucklin, Kansas, the *Madison Journal* of Tallulah, Louisiana, and the *Winston County Journal* of Louisville, Mississippi.

2. The February 4, 1913, edition of the Texas *Amarillo Daily News* treated the event as if it had already been commonly discussed by its readers noting simply and cryptically in its Top o' the Morning section that "Battles between a bull and elephant in the Juarez arena, shows that American political methods have crossed the border." For a copy of the ad, see Tusko the elephant, news clippings, 1933, 152, Woodland Park Zoo Historical and Administrative Records, Record Series 8601-01, box 15, file 2, Seattle Municipal Archives.

3. "Bull in a Fight with an Elephant," *Lyons (KA) Republican*, June 3, 1913.

4. See Homer C. Walton, "The M. L. Clark Wagon Show," *Bandwagon* 9, no. 2 (1965): 4–11.

5. This story is similar, in fact, to the one related in Silvio Bedini's *The Pope's Elephant* (Manchester, UK: Carcanet, 1997) of a staged battle between an elephant and a rhinoceros at the beginning of the sixteenth century. Many of the features of this story echo, too, a staged fight between an American buffalo and bull that took place in the same Juárez arena in 1907. According to an Albuquerque paper, the crashes between those two animals could be "heard for blocks." See "Big Buffalo Defeats Bull Sunday: In the Ring at Juarez, before an Immense Crowd of People," *Albuquerque (NM) Evening Citizen*, January 29, 1907.

6. The name Ruhe may have been attached to the history of Ned simply because the company was famous in its own right.

7. I am pleased to have the opportunity to thank Richard J. Reynolds III for his lively correspondence and enthusiastic support over now many years. When we started talking about Tusko in January 2019, he quickly sent off an excerpt about Tusko from his forty-year-old paper that he presented at the 1979 Knoxville, Tennessee, Regional Workshop of the American Association of Zoological Parks and Aquarium, entitled "Hold Your Horses, Here Come the Elephants!" That he would have quick access to something he wrote forty years ago, that it would have such a title, and that he would share it so generously, is all very typical of Richard.

8. Walton, "The M. L. Clark Wagon Show."

9. Other sources list Mena as having arrived in the United States in 1890, but the 1895 date seems more plausible. In the spring of 2019, I asked the archivist at the Hagenbeck Tierpark in Hamburg to see if the Clark Circus was listed as a buyer in the

1890s. The records are incomplete, but he could not find any documentation of a sale to the circus.

10. M. L. Clark and Son's picked up two more quite young elephants, Tony and Babe, in 1908 but initially they rode in a wagon. Tony was sold to Al G. Barnes in 1909.

11. The camel is possibly one of the six camels including the locally famous Mose that Clark picked up at the end of the 1904 World's Fair in St. Louis. See Homer C. Walton, "Ned and Mena, Famous Elephants," *Bandwagon* 2, no. 6 (1958): 7.

12. Walton, "Ned and Mena, Famous Elephants," 7.

13. See, for example, the report on the Duggan Bros. Circus in the *Daily Reporter* (Greenfield, IN), July 16, 1934, the report on Bailey Bros. Circus in the *Moberly (MO) Monitor-Index*, June 6, 1935, and the report on the Johnny J. Jones Exposition in the *Florence (SC) Morning News* on October 8, 1935.

14. *Daily Reporter* (Greenfield, Indiana), July 12, 1934.

15. Homer Walton makes a similar observation: "While on the Clark Show Ned did not get so much of a name as a bad elephant than he did later on. But, this was probably because walking from town to town over the tough roads in the south, pushing wagons out of mud holes and off and on muddy lots as well as working in the performances each day, helped to keep his mind off trouble" ("Ned and Mena, Famous Elephants," 7). "Buckles" Woodcock has returned several times over the years to the relatively well-known photograph of Ned and Mena on his Buckles blog. See, for example, his post and follow-up comments of October 11, 2005, https://bucklesw.blogspot.com/2005/10/ml-clark-circus-c1919-ned-and-mena.html.

16. Richard J. Reynolds, "Hold Your Horses, Here Come the Elephants!" Reynolds's date, location, and destination accord with Walton's account in "The M. L. Clark Wagon Show." According to Alexander Haufellner, Jürgen Schilfarth, and Georg Schweiger in *Elefanten in Zoo und Circus*, vol. 2: *200 Jahre Elefantenhaltung in Nordamerika 1796–1996* (Munster: Schüling, 1997), however, Ned was only on the Clark Circus until 1916, after which he was transferred to a series of other circuses, including the Wheeler Bros. Circus and the R. T. Richard Shows. I have not been able to corroborate these transfers through any other accounts of Ned nor through accounts of those circuses. For these reasons, I am sticking to the story that has been told of Ned for a very long time—that the Clark Circus sold him to Barnes in 1921.

17. Reynolds, "Hold Your Horses, Here Come the Elephants!"

18. Tusko was, in the end, carefully measured by a circus fans group in 1932, and his numbers were an impressive 10 feet, two inches high and 14,313 pounds. One should note, though, that that weight came after several years during which the elephant's circumstances were rather borderline and his keepers struggled to afford to feed him. I think it likely that Ned weighed substantially more when he was on the Barnes circus, though obviously not twenty-thousand pounds.

19. "What Happened When Tusko Went on a Rampage," as told to Dave Roberson by Al G. Barnes, typescript, 1. Woodland Park Zoo Historical and Administrative Records, Record Series 8601-01, box 15, folder 3, Seattle Municipal Archives.

20. Buckles blog is one of the great sources on the history of elephants in circuses. This post was from October 23, 2016; see https://bucklesw.blogspot.com/2016/10/7-tusko.html.

21. "What Happened When Tusko Went on a Rampage," 3.

22. "Elephant on the Rampage: He Leaves Thirty-Mile Trail of Destruction in Washington State," *New York Times*, May 18, 1922.

23. There was no one named "Hendrickson" associated with the Barnes Circus in 1922. The name does not appear in the *Official Season Route Book and Itinerary*, which lists all the circus staff for the year, along with the dates of all the performances. A copy of the 1922 route book is in the collection of the Robert L. Parkinson Library and Research Center at the Circus World Museum Library in Baraboo, Wisconsin.

24. "What Happened When the Elephant 'Took a Notion,'" *Salt Lake Telegram*, magazine section, 29.

25. The ad men for the Barnes Circus seem to have been active in placing stories in local newspapers across the country, although it can be difficult to trace them. Essentially the same article (with local details changed) appeared in newspapers across the country. The articles would often be published in the afternoon edition of the newspaper on the day the circus was in town and would encourage people to attend the evening performance. When the circus was in Visalia, California, on March 30, 1922, for example, an article titled "Al G. Barnes Circus Here; Parade, First Show Please" that described the parade and the afternoon performance appeared in the local paper. Essentially the same article appeared on May 5, 1922, in the *Eugene (OR) Guard* as "Circus Parade Delights Kids, Older Folks," and on October 18, 1922 in the *Corsicana (TX) Daily Sun* as "Great Crowds Delight in Al G. Barnes Circus Parade."

26. "Elephant Star of Parade: Tusko a Drawing Card for Al G. Barnes' Circus That Shows Monday Evening," *Lincoln (NE) Journal Star*, June 26, 1922.

27. "'Tusko,' of the Circus Escapes and Is Caught on P.R.R. Tracks," *Evening News* (Harrisburg, PA), August 9, 1922.

28. The first page of the ship's logbook notes the departure of the ship from Calcutta on December 3, 1795, a little over four months earlier. In his 1925 account of the log and the elephant, George Gilbert Goodwin, curator of mammals at the American Museum of Natural History, notes that the log first mentions the elephant on February 17, two months after the departure, while the ship was taking on supplies at the island of St. Helena in the south Atlantic. Two months later, the elephant was unloaded in New York ("The First Living Elephant in America," *Journal of Mammalogy* 6, no. 4 [1925]: 256–63).

29. Goodwin, "The First Living Elephant in America," 259.

30. It is not clear how long this elephant survived. An elephant that became known as "Old Bet" and that was purchased by Hachaliah Bailey in 1808 may well have been the Crowninshield elephant. Bailey's "Old Bet" appears to have died in 1816.

31. The broadsheet is reproduced in Goodwin, "The First Living Elephant in America," pl. 24

32. Elizabeth Sandwith Drinker, *Diary of Elizabeth Sandwith Drinker*, vol. 2 (Boston: Northeastern University Press, 1991), 860.

33. I have written about a "one-foot stand" in "Why Look at Elephants?" *Worldviews Environment, Culture, Religion* 9, no. 2 (2005): 166–83.

34. "What Happened When Tusko Went on a Rampage," 3.

35. The figure of twenty-five cents comes from a typescript titled "History of Tusko" prepared by Gus Knudson, director of the Woodland Park Zoo, in the fall of

1932. Knudson attempted to trace Tusko's past through interviews and correspondence when the zoo acquired the elephant in October of that year. Many of the dates in the account are incorrect (for example, he gives date of the transfer to Barnes as 1922 and date of the Sedro-Woolley "rampage" as 1924), but the general outline is good (Woodland Park Zoo Historical and Administrative Records, Record Series 8601-01, box 15, folder 3, Seattle Municipal Archives).

36. "What Happened When Tusko Went on a Rampage," 4.

37. I use the word "pen" partly because of a note on the back of a photograph at the Circus World Museum of Tusko from 1930 when he was in the Barnes new winter quarters, Baldwin Park. A similar pen was set up for Tusko there, but with a wood barn built around it. In the pen was a stall with iron on all four sides that could essentially immobilize Tusko completely. On the back of the photograph which shows Tusko in the stall, someone has written "Tusko in his first jail, harness just being fitted or tried on rather Culver City. Mr Forbes on top of elephant who designed harness and jail." "Forbes" is Red Forbes, the blacksmith with the Barnes show who also designed the chains for Tusko.

38. When the circus was sold, it continued to travel into the 1930s as the Al G. Barnes Circus. In 1928 and 1929, Tusko was shown beside another big Asian bull named Diamond, acquired by Barnes in 1928 as a stand-in for Tusko when the elephant was too dangerous to be with the show. Thus, when Tusko was in his pen in California, Diamond (usually referred to now as Black Diamond, the name given him by William Woodcock Sr.) would be called Tusko, and if they were both present at a show Diamond would be called Tusko and Tusko would be called Mighty Tusko. They were both with the circus in Corsicana, Texas, on October 12, 1929, when Diamond killed a local woman, Eva Speed Donohoo. See Homer C. Walton, "The Story of Black Diamond," *Bandwagon* 3, no. 3 (1959): 17–18.

39. "Running amok" like "rogue," seems to be an expression inextricably tied to the Western encounter with India, Malaysia, and elephants. I write about Hero in "A Hero's Death," *Animal Acts: Performing Species Now*, ed. Una Chaudhuri and Holly Hughes (Ann Arbor: University of Michigan Press, 2013), 182–88, and about Diamond in "Touching Animals: The Search for a 'Deeper Understanding' of Animals," in *Beastly Natures: Animals, Humans, and the Study of History*, ed. Dorothee Brantz (Charlottesville: University of Virginia Press, 2010), 38–58.

40. The Circus Historical Society has a large data project on circus route books. These dates come from the database; see https://circushistory.org/archive/routes.

41. "Elephant Derby Next Item on Sports List," *Oakland (CA) Tribune*, December 21, 1931.

42. See "Tusko Is Sold," *Bend (OR) Bulletin*, November 5, 1931. George "Slim" Lewis claimed in his memoirs (cowritten with Byron Fish), *I Loved Rogues: The Life of an Elephant Tramp* (Seattle: Superior Publishing, 1978), that O'Grady and Gray managed to get the title for a dollar (117).

43. Lewis, *I Loved Rogues*, 118.

44. "Owners Make New Home for Huge Elephant," *Salt Lake Tribune*, November 29, 1931.

45. See "Tusko Sale Falls Through," *Bend (OR) Bulletin*, December 9, 1931, and "World's Biggest Toddy Given Tusko," *Victoria (TX) Advocate*, December 21, 1931.

46. "Tusko on Rampage, Again: Dodges Firing Squad," *Press Democrat* (Santa Rosa, CA), December 26, 1931; "Big Elephant Rechained in Wild Battle," *Salt Lake Telegram*, December 26, 1931; "Huge Animal Saved from Firing Squad: Elephant Goes Berserk, but Mayor Steps in to Halt Guns and Quiet Beast," *Abilene (TX) Reporter-News*, December 27, 1931.

47. An article entitled "Elephant Business Booms as Six Tons of Tusko Are Viewed by Portland People" in the *Klamath (OR) News* of December 27, 1931, claimed that six hundred people paid to see Tusko the day after the "rampage" and that "Tusko, Elephant, Unltd." was back to doing impressive business.

48. Lewis, *I Loved Rogues*, 129.

49. "Salem May Send Elephant Abroad: Citizens of Capital City Start Fund to Send Tusko to Siam," *Eugene (OR) Guard*, November 4, 1931.

50. "Was It a Mercy?," in the "What Other Editors Think" column in the *Eugene (OR) Guard*, December 30, 1931.

51. Rose Hellman to Gus Knudson, October 9, 1932, and November 25, 1932. Woodland Park Zoo Historical and Administrative Records, Record Series 8601-01, box 15, folder 4, Seattle Municipal Archives.

52. Lewis, *I Loved Rogues*, 143.

53. Report on Tusko, April 1933, Woodland Park Zoo Historical and Administrative Records, Record Series 8601-01, box 15, folder 8, Seattle Municipal Archives.

54. George W. Lewis, report on Tusko, Woodland Park Zoo Historical and Administrative Records, Record Series 8601-01, box 15, folder 8, Seattle Municipal Archives. Reports by other keepers and officials of the zoo corroborate Lewis's account.

55. Christen Wemmer and Catherine A. Christen, eds., *Elephants and Ethics: Toward a Morality of Coexistence* (Baltimore, MD: Johns Hopkins University Press, 2008). My thanks to Chris and Kate for including me in this remarkable project.

56. For more on Keiko, see Jason Colby, *Orca: How We Came to Know and Love the Ocean's Greatest Predator* (New York: Oxford University Press, 2018).

57. It is worth remembering that as much as the circus industry has used elephants to magnify their entertainments, People for the Ethical Treatment of Animals (PETA)—along with zoos, tourist bureaus, and all kinds of companies—have used them to magnify their messages as well.

Chapter 6 • The Last of Its Kind

1. George Peress Sanderson, *Thirteen Years among the Wild Beasts of India*, 2nd. ed. (London: William H. Allen, 1879), 2.

2. While Sanderson admired Baker, he was critical of Tennent's work, calling it "full of the errors which are unavoidable when a man writes on a subject with which he has no practical acquaintance" (*Thirteen Years among the Wild Beasts of India*, 65). While it is true that Tennent seems to have often relied on stories he had been told about elephants, Sanderson appears to have been more frustrated by Tennent's overall characterization of elephants as peaceful, timid, inoffensive, and retiring creatures who, in reality, presented no real challenge for a hunter: "Sir Emerson Tennent being confessedly no sportsman probably never saw a wild tusker" and "never in his life encountered elephants when roused to anger, which must be taken into consideration in accepting his view of the matter" (193).

3. Sanderson, *Thirteen Years among the Wild Beasts of India*, 189.

4. Sanderson notes that "it is certain that in a few years the interdiction will have to be relaxed, as elephants are being preserved without corresponding measures being taken for their reduction by capture" (*Thirteen Years among the Wild Beasts of India*, 187).

5. See Carl Georg Schillings, *Mit Blitzlicht und Büchse: Neue Beobachtungen und Erlebnisse in der Wildnis inmitten der Tierwelt Äquatorial-Ostafrika* (Leipzig: Voigtländer, 1904).

6. Theodore Roosevelt, *Outdoor Pastimes of an American Hunter* (New York: Scribner's, 1905), 336.

7. Roosevelt, *Outdoor Pastimes*, 328.

8. For more on Schillings and his ideas, especially in Germany, see Bernhard Gissibl, *The Nature of German Imperialism: Conservation and the Politics of Wildlife in Colonial East Africa* (New York: Berghahn, 2016), especially 270–78.

9. Theodore Roosevelt to William T. Hornaday, January 17, 1906, Theodore Roosevelt Papers, Library of Congress Manuscript Division, for a reproduction of the letter, see Theodore Roosevelt Digital Library. Dickinson State University, https://www.theodorerooseveltcenter.org/Research/Digital-Library/Record?libID=o194007.

10. Schillings' descriptions of landscapes often focus on the strange visual effects of light, as well: "Darker spots in the distance far away from us we take to be larger wild animals. The field-glass shows that they are hartebeests, and a great number of waterbuck; and still farther off there is a moving mass that shimmers and is half lost in the glare of the morning sun. There are zebras, and yet more zebras, moving like living walls! Strange effects of light actually give us the impression of something like a wall or rampart, made up of the living forms of the zebras—the deep shadows they throw come out black, their flanks are lighted up in the dazzling sunshine, and they shimmer with all colours and with ever-changing effect" (*In Wildest Africa*, trans. Frederic Whyte [New York: Harper, 1907], 20).

11. A digitized version of the 1906 edition of *With Flashlight and Rifle* is available from archive.org: https://archive.org/details/withflashlightri01schiiala. The copy from which the scan was made is housed at the University of California, Santa Barbara, and originated in the personal library of another notable writer about Africa, Elspeth Huxley. The cover is based on a photograph captioned "The Huge Elephant Fell Dead," reproduced on p. 189 of the edition. The elephant collapsed in an "upright" position—not on its side.

12. Carl Georg Schillings, *With Flashlight and Rifle: A Record of Hunting Adventures and of Studies in Wild Life in Equatorial East Africa*, trans. Frederic Whyte (London: Hutchinson, 1906), 1–2.

13. Schillings, *With Flashlight and Rifle*, 4–5.

14. Schillings, *With Flashlight and Rifle*, 388, 736. In an appendix to Schillings' account titled "A Few Words about Herr C. G. Schillings' Collection of East African Mammals," the Berlin taxonomist Paul Matschie notes that beyond bringing specimens of over one hundred different species of mammals, Schillings usually brought many, many specimens of each species. He points out that the hunter "brought home quite a number of giraffes, buffaloes, rhinoceroses, and elephants, a great number of large antelopes, and hundreds of hides and skins and skeletons of every description, all of them in such good condition that they are suitable for exhibition in museums" (730).

15. Schillings, *With Flashlight and Rifle*, 741.
16. Carl Georg, *Der Zauber des Elelescho* (Leipzig: Voigtländer, 1906).
17. Schillings, *In Wildest Africa*, 1.
18. Schillings, *In Wildest Africa*, 19.
19. Schillings, *In Wildest Africa*, 44.
20. Schillings, *In Wildest Africa*, 48.
21. Schillings, *In Wildest Africa*, 56.
22. Sometimes called "symphonic poems," tone poems were related to multimovement program symphonies such as Berlioz's *Symphonie fantastique* (1830). Key composers include Franz Liszt, Claude Debussy, Jean Sibelius, and Richard Strauss. Some of the better-known works today are Strauss's 1915 *An Alpine Symphonie*, Paul Dukas's 1897 *The Sorcerer's Apprentice*, based on a poem by Goethe, and Modest Mussorgsky's 1867 *Night on Bald Mountain*, the latter two of which were included in Disney's *Fantasia*.
23. Schillings, *In Wildest Africa*, 52–61.
24. Schillings, *In Wildest Africa*, 114.
25. Schillings, *In Wildest Africa*, 516.
26. Schillings, *In Wildest Africa*, 519.
27. Schillings, *In Wildest Africa*, 536.
28. In his appendix to *With Flashlight and Rifle*, Matschie makes clear that from his scientific perspective as well, hunters like Schillings were simply not the problem. Noting that critics erroneously accused Schillings of "having helped to exterminate wild life in the regions which he had visited" (735–36), Matschie insists that Schillings had always been solely committed to increasing knowledge. Indeed, Matschie writes, Schillings had done so much for science that several new species, including *Giraffa schillingsi*, *Hyena schillingsi*, and a new species of klipspringer designated as *Oreotragus schillingsi*, had been named in his honor (741). These names have since been dropped, but a kind of tick (*Ixodes schillingsi*, Neumann, 1901), a longhorn beetle (*Prionotoma schillingsi*, Lameere, 1903), a biting fly (*Haematobia schillingsi*, Grunberg, 1906), and a subspecies of cisticola warbler (*Cisticola cinereolus schillingsi*, Reichenow, 1905) still carry Schillings' name.
29. Schillings, *With Flashlight and Rifle*, 194.
30. Schillings, *With Flashlight and Rifle*, 198.
31. Schillings called the site Kilepo Hill, the "watering place" in the Maasai or Maa language.
32. Schillings, *In Wildest Africa*, 541.
33. Schillings, *In Wildest Africa*, 542.
34. Schillings, *In Wildest Africa*, 547.
35. Schillings, *In Wildest Africa*, 549.
36. Schillings, *In Wildest Africa*, 572–77.
37. See Mark V. Barrow, *Nature's Ghosts: Confronting Extinction from the Age of Jefferson to the Age of Ecology* (Chicago: University of Chicago Press, 2009).
38. Charles Lyell, *Principles of Geology, Being an Attempt to Explain the Former Changes of the Earth's Surface, by Reference to Causes Now in Operation*, vol. 2 (London: Murray, 1832), 155–56. Man was not, however, the only cause of extinction. Lyell writes, "The most insignificant and diminutive species have each slaughtered their thousands, as they disseminated themselves over the globe" (156).

39. William T. Hornaday, "National Collection of Heads and Horns," *Zoological Society Bulletin* 40 (1910): 667.

40. Hornaday, "National Collection of Heads and Horns," 668.

41. I discuss Akeley's expedition to Somaliland in "Trophies and Taxidermy," in *Gorgeous Beasts: Animal Bodies in Historical Perspective*, ed. Joan Landes, Paula Young Lee, and Paul Youngquist (State College: Penn State University Press, 2012), 117–36.

42. Two of the elephants he and his wife collected became his gigantic work of 1909, *The Fighting Bulls*, which can still be seen in the main hall of the Field Museum. For more on *The Fighting Bulls*, see Nigel Rothfels, "Preserving History: Collecting and Displaying in Carl Akeley's *In Brightest Africa*," in *Animals on Display: The Creaturely in Museums, Zoos, and Natural History*, ed. Karen Rader, Liv Emma Thorsen, and Adam Dodd (State College: Penn State University Press, 2013), 58–73.

43. Carl E. Akeley, *In Brightest Africa* (Garden City, NY: Garden City Publishing, 1920), 23–24.

44. Akeley, *In Brightest Africa*, 26.

45. Akeley, *In Brightest Africa*, 27.

46. Carl E. Akeley, "Elephant Hunting in Equatorial Africa with Rifle and Camera," *National Geographic Magazine*, August 1912, 797.

47. Akeley, *In Brightest Africa*, 55.

48. See the description of this hunt in Darrin Lunde, *The Naturalist: Theodore Roosevelt, A Lifetime of Exploration, and the Triumph of American Natural History* (New York: Broadway Books, 2016), 240–43. Four additional elephants were added to the group in the 1930s—after Akeley's death—to expand the work to the eight elephants that can be seen today. These four elephants were collected by F. Trubee Davidson and his wife, Dorothy Peabody, in 1933. See Steven Christopher Quinn, *Windows on Nature: The Great Habitat Dioramas of the American Museum of Natural History* (New York: Abrams and American Museum of Natural History, 2006).

49. Akeley, *In Brightest Africa*, 55.

50. Akeley, "Elephant Hunting in Equatorial Africa with Rifle and Camera," 810.

51. T. J. Chapman, "Doomed," *Forest and Stream: A Weekly Journal of Rod and Gun* 59, no. 2 (1902): 22.

52. Chapman, "Doomed"; see also Frank Vincent, *Actual Africa: Or, the Coming Continent* (New York: Appleton, 1895), 472.

53. Chapman, "Doomed."

54. Arthur H. Neumann, *Elephant Hunting in East Equatorial Africa: Being an Account of Three Years' Ivory-Hunting under Mount Kenia and among the Ndorobo Savages of the Lorogi Mountains, including a Trip to the North End of Lake Rudolph* (London: Rowland Ward, 1898), viii.

55. Alfred Edmund Brehm, *Brehms Thierleben: Allgemeine Kunde des Thierreichs*, 2nd ed., vol. 3 (Leipzig: Verlag des Bibliographischen Instituts, 1877), 501, my translation.

Chapter 7 • Trails of History

1. Bertolt Brecht, *Gesammelte Werke*, vol. 12: *Prosa 2* (Frankfurt: Suhrkamp, 1967), 387–88, my translation. My thanks to Stephan Oettermann for leading me to this remarkable quotation.

2. Jonathan Swift, *On Poetry: A Rhapsody*, in *The Poems of Jonathan Swift*, vol. 2, ed. Harold Williams (Oxford, UK: Clarendon Press, 1937), 645–46.

3. See Matthew Edney, "A Misunderstood Quatrain," December 15, 2018, *Mapping as Process* blog, https://www.mappingasprocess.net/blog/2018/12/15/a-misunderstood-quatrain. An updated version of Swift might be Captain Darling in the 1980s BBC comedy series *Blackadder* asking incredulously about a map Blackadder claimed to have prepared of No-Man's Land in the trenches of World War I: "Are you sure this is what you saw, Blackadder?" Blackadder responds: "Absolutely. I mean, there may have been a few more armament factories and not quite as many elephants." See *Blackadder Goes Forth*, season 4, episode 1, "Captain Cook," aired September 28, 1989.

4. Carl E. Akeley, *In Brightest Africa* (Garden City, NY: Garden City Publishing, 1920), 35.

5. Akeley, *In Brightest Africa*, 35–36.

6. My thanks to Carl Bogner for introducing me to the unusual documentary film of a lecture delivered by Raymond L. Birdwhistell to the American Anthropological Association in 1966 that is also a trail titled *Microcultural Incidents in Ten Zoos: An Illustrated Lecture*. The film presents a cross-cultural comparison of different people from around the world standing before elephant exhibits.

7. Gordon G. Gallup Jr., "Chimpanzees: Self Recognition," *Science* 167 (1970): 86–87.

8. Gallup, "Chimpanzees," 87.

9. Gallup, "Chimpanzees," 87.

10. On March 27, 2020, Ambika was euthanized; she was reckoned to be about seventy-two years old. At that time, Shanthi, at forty-five years old, continued to live at the National Zoo.

11. Daniel J. Povinelli, "Failure to Find Self-Recognition in Asian Elephants (*Elephas maximus*) in Contrast to Their Use of Mirror Cues to Discover Hidden Food," *Journal of Comparative Psychology* 103, no. 2 (1989): 130.

12. Povinelli, "Failure to Find Self-Recognition in Asian Elephants (*Elephas maximus*) in Contrast to Their Use of Mirror Cues to Discover Hidden Food," 130.

13. Sammy (also known as Maya, Sammi, Sami, and Sammy R) was born in captivity on April 17, 1992, at Busch Gardens in Tampa, Florida, and was euthanized on January 31, 2006, at fourteen years of age, because he was apparently suffering from liver disease. "Self-Recognition in an Asian Elephant" was submitted to PNAS on September 13, 2006, and published on November 7, 2006. The article did not note the death of the young elephant. In the fall of 2018, Maxine died; Patty and Happy remain at the Bronx Zoo.

14. The fact that Happy was not interested in the mark on subsequent days might not be significant. Primates also seem to lose interest in the mark in subsequent experiments. Magpies, however, do remain interested in the marks, and researchers hypothesize that the condition of their feathers might be more important to magpies than a mark on the skin to elephants or chimpanzees. See Helmut Prior, Ariane Schwarz, and Onur Güntürkün, "Mirror-Induced Behavior in the Magpie (*Pica pica*): Evidence of Self-Recognition," *PLoS Biology* 6, no. 8 (2008): e202.

15. Joshua M. Plotnik, Frans B. M. de Waal, and Diana Reiss, "Self-Recognition in an Asian Elephant," *Proceedings of the National Academy of Sciences* 103, no. 45 (2006): 17055.

16. Plotnik, de Waal, and Reiss, "Self-Recognition in an Asian Elephant," 17055.

17. "First Evidence to Show Elephants Recognize Themselves in The Mirror," Emory University Press Release, October 30, 2006, http://whsc.emory.edu/press_releases_print.cfm?announcement_id_seq=8080.

18. Andrew Bridges, "Mirror Test Implies Elephants Self-Aware," *Associated Press*, October 30, 2006, https://www.washingtonpost.com/wp-dyn/content/article/2006/10/30/AR2006103000881.html?noredirect=on.

19. I include bats in this list as a nod to Thomas Nagel's seminal essay "What Is It Like to Be a Bat?," *Philosophical Review* 83, no. 4 (1974): 435–50.

20. Richard W. Byrne and Lucy Bates, "Elephant Cognition: What We Know about What Elephants Know," in *The Amboseli Elephants: A Long-Term Perspective on a Long-Lived Mammal*, ed. Cynthia J. Moss, Harvey Croze, and Phyllis C. Lee (Chicago: University of Chicago Press, 2011), 181–82.

21. It is worth saying that to this day one of the very oldest representational figurines that has ever been found is a small, inch-and-a-half long carving of a mammoth made from mammoth ivory over thirty-thousand years ago. The little sculpture was found in 2007 in dirt associated with the so-called Vogelherd Cave in what is now southwest Germany.

22. I write more about mammoths and Knight in "Mammoths in the Landscape," in *Routledge Handbook of Human-Animal Studies*, ed. Susan McHugh and Garry Marvin (London: Routledge, 2014), 10–22. See also, Charles R. Knight, *Prehistoric Man: The Great Adventurer* (New York: Appleton-Century-Crofts, 1949).

23. Charles Siebert, "An Elephant Crackup?" *New York Times Magazine*, October 8 2006, https://www.nytimes.com/2006/10/08/magazine/08elephant.html. Siebert's argument was substantially based on the thoughtful research of the advocate for trans-species psychology, Gay Bradshaw. See G. A. Bradshaw, *Elephants on the Edge: What Animals Teach Us About Humanity* (New Haven, CT: Yale University Press, 2009).

24. Raman Sukumar, *The Living Elephants: Evolutionary Ecology, Behaviour, and Conservation* (Oxford: Oxford University Press, 2003). See also *Conflict, Negotiation, and Coexistence: Rethinking Human-Elephant Relations in South Asia*, ed. Piers Locke and Jane Buckingham (New Delhi: Oxford University Press, 2016).

25. Lumpers and splitters continue discussion of how many kinds, or species, of extinct proboscideans there are. In the 1930s and '40s, Henry Fairfield Osborn listed around 350 species; recent lists included just over 150. For a discussion, see Sukumar's *The Living Elephants*, 3–45.

26. My deep thanks to Elizabeth Frank for all her friendship, encouragement, and guidance over the last twenty-five years. My thanks, too, to Charles Wikenhauser and Tracey Dolphin-Drees for helping me learn so much more about zoos and elephants early on in this project.

27. Jeffrey Moussaieff Masson and Susan McCarthy, *When Elephants Weep: The Emotional Lives of Animals* (New York: Delta, 1995).

28. Charles Darwin, *The Expression of the Emotions in Man and Animals* (London: John Murray, 1872), 167–68.

29. James Emerson Tennent, *Ceylon: An Account of the Island Physical, Historical, and Topographical with Notices of Its Natural History, Antiquities and Productions* (London: Longman, 1859), 376.

30. In the original maquette (complete and assembled draft) of the story of Babar, the book's first two plates were of the hunting and killing of Babar's mother. In the final published version, three plates were added showing Babar in domestic bliss before the jarring hunting scene. The image of Babar crying next to his dead mother has always reminded me of the photograph by Hans Schomburgk of his elephant Jumbo standing beside his dead mother. See Christine Nelson, *Drawing Babar: Early Drafts and Watercolors* (New York: Morgan Library, 2008), 31–32.

31. Pliny, *The Natural History of Pliny*, vol. 2, trans. Harris Rackham (Cambridge, MA: Harvard University Press, 1942), 509.

32. Aelian, *On the Characteristics of Animals*, vol. 2, trans. A. F. Scholfield (Cambridge, MA: Harvard University Press, 1958), 309.

33. Edward Topsell, *The History of Four-Footed Beasts, Serpents, and Insects* (London: Cotes, 1658), 154.

34. Georges-Louis Leclerc de Buffon, *Natural History: General and Particular by the Count de Buffon*, trans. William Smellie, 2nd ed., vol. 6 (London: Strahan and Cadell, 1785), 7–8.

35. Masson and McCarthy, *When Elephants Weep*, 109.

36. There has been a long interest in whether elephants can weep emotional tears. See Robert Harrison, "On the Anatomy of the 'Lachrymal Apparatus' in the Elephant," *Proceedings of the Royal Irish Academy* 4 (1848): 158–65, Morrison Watson, "Contributions to the Anatomy of the Indian Elephant," pt. 3, "The Head," *Journal of Anatomy and Physiology* 8, no. 1 (1873): 85–94, and E. T. Collins, "The Physiology of Weeping," *British Journal of Ophthalmology* 16, no. 1 (1932): 1–20. More recent studies confirm that elephants lack tear ducts and are physically unable to "cry" in the way humans do.

37. Marc Bekoff, "Animal Emotions: Exploring Passionate Natures," *BioScience* 50, no. 10 (2000): 868.

38. Aelian, *On the Characteristics of Animals*, 95–97.

39. Aelian, *On the Characteristics of Animals*, 75.

40. Polybius, *Histories*, vol. 2, trans. Evelyn Shirley Shuckburgh (London: Macmillan, 1889), 144.

A is for Alice who lived in a zoo. The first elephant mentioned in this book is Alice, who was born in India in the 1890s, brought to Germany by Carl Hagenbeck, and then delivered to Luna Park in the spring of 1904 by Hagenbeck's son Lorenz. In his memoir, *Animals Are My Life* (London: Bodley Head, 1956), Lorenz Hagenbeck recalls his experience of bringing twenty elephants to Luna Park in this single shipment—a shipment that actually included another sixteen elephants destined for other buyers (47–49). Traveling with Alice to Luna Park was another elephant who came to be known as Princess Alice. Within a couple years she was sold to a circus and eventually ended up in Salt Lake City's Hogle Zoo, where she died in 1953. When I was growing up in Salt Lake, I used to look up at a relief sculpture of Princess Alice on the old elephant building at the zoo—the place to which I can trace some of my earliest memories of elephants. The Bronx's Alice, the Alice whom Helen Keller met and Gunda knew, lived at the Bronx Zoo for thirty-five years and died in 1943.

Once you start looking for the presence of elephants in our histories, you begin to find them everywhere. The database of over fifteen thousand elephants that have lived or are living in captivity created by Dan Koehl (https://www.elephant.se) currently lists twenty-eight elephants named Alice. There is a glacial erratic named the Stone Elephant on a trail where I sometimes walk. There is a giant, pink fiberglass elephant wearing glasses in DeForest, Wisconsin, that I see sometimes when I am headed out on camping trips. Centrally, this book argues that elephants have occupied human thoughts more than most other animals for a very long time, and so it should not be surprising that elephant kitsch abounds, that I have been given many elephant-themed ties and trinkets, and that the existing literature about elephants is vast.

In thinking about recommendations for further reading, I want to draw attention to works that have helped me develop my ideas for this project over the last twenty years. I especially recommend Iain and Oria Douglas-Hamilton's *Among the Elephants* (New York: Viking, 1975), Cynthia Moss's *Elephant Memories: Thirteen Years in the Life of an Elephant Family* (1988; rpt., Chicago: University of Chicago Press, 2000), and Raman Sukumar's *The Living Elephants: Evolutionary Ecology, Behaviour, and Conservation* (Oxford: Oxford University Press, 2003). For accounts of human history and elephants, I recommend Silvio Bedini's *The Pope's Elephant: An Elephant's Journey from Deep in India to the Heart of Rome* (New York: Penguin, 1997), *Elephants and Ethics: Toward a Morality of Coexistence*, edited by Christen Wemmer and Catherine A. Christen (Baltimore, MD: Johns Hopkins, University Press, 2008), Dan Wylie's *Elephant* (London: Reaktion, 2012), and Stephan Oettermann's *Die Schaulust am Elefanten: Eine Elephantographia Curiosa* (Frankfurt: Syndicat, 1982), a book I first read more than thirty-five years ago that takes its title from Georg Christoph Petri von Hartenfels's *Elephantographia curiosa* (Leipzig, 1715). I also recommend Barbara Gowdy's unusual and compelling fictional work about elephants, *The White Bone: A Novel* (Toronto: HarperCollins Canada, 1998).

I found myself often turning to older works of natural history as I wrote this book. *The Natural History of Pliny*, translated by John Bostock and H. T. Riley (London: Henry Bohn, 1855), Aelian's *On the Characteristics of Animals*, translated by A. F. Scholfield (Cambridge, MA: Harvard University Press, 1958), Edward Topsell's *The History of Four-Footed Beasts, Serpents, and Insects* (London: Cotes, 1658), Buffon's *Natural History: General and Particular by the Count de Buffon*, translated by William Smellie (London: Strahan and Cadell, 1785), and the first two editions of Alfred Brehm's *Animal Life—Illustrirtes Thierleben* (1864–69) and *Brehms Thierleben: Allgemeine Kunde des Thierreichs* (Leipzig: Bibliographisches Institut, 1876–79)—have been particularly important to me. In addition, I want to draw attention to Howard Hayes Scullard's *The Elephant in the Greek and Roman World* (London: Thames and Hudson, 1974), and George Druce's "The Elephant in Medieval Legend and Art," in the *Journal of the Royal Archaeological Institute* 76 (1919): 1–73.

In the second volume of his *Ceylon: An Account of the Island Physical, Historical, and Topographical with Notices of Its Natural History, Antiquities and Productions* (London: Longman, 1859), James Emerson Tennent observes that the hunting of elephants "has been described in tiresome iteration" (326). My shelves have been especially weighed down with nineteenth-century works on hunting and exploration. The most important for me have been William Cornwallis Harris's *Wild Sports of Southern Africa* (London: John Murray, 1839), Roualeyn Gordon-Cumming's *Five Years of a Hunter's Life in the Far Interior of South Africa* (New York: Harper Brothers, 1850), Samuel White Baker's *The Rifle and the Hound in Ceylon* (London: Longman, 1854), Arthur H. Neumann's *Elephant Hunting in East Equatorial Africa* (London: Rowland Ward, 1898), George P. Sanderson's *Thirteen Years among the Wild Beasts of India* (London: Allen, 1879), Hans Hermann Schomburgk's *Wild und Wilde im Herzen Afrikas; Zwölf Jahre Jagd- und Forschungsreisen* (Berlin: Fleischel, 1910), Carl Georg Schillings' *Mit Blitzlicht und Büchse: Neue Beobachtungen und Erlebnisse in der Wildnis inmitten der Tierwelt Äquatorial-Ostafrika* (Leipzig: Voigtländer, 1904) and his *Der Zauber des Elelescho* (Leipzig: Voigtländer, 1906), Theodore Roosevelt's *African Game Trails: An Account of the African Wanderings of an American Hunter-Naturalist* (London: John Murray, 1910), and Carl E. Akeley's *In Brightest Africa* (Garden City, NY: Garden City Publishing, 1920). For their insightful readings about the history of hunting, I recommend Matt Cartmill's *A View to a Death in the Morning: Hunting and Nature through History* (Cambridge, MA: Harvard University Press,

1993), Bernhard Gissibl's *The Nature of German Imperialism: Conservation and the Politics of Wildlife in Colonial East Africa* (New York: Berghahn, 2016), Andrew C. Isenberg's *The Destruction of the Bison* (Cambridge: Cambridge University Press, 2000), Darrin Lunde's *The Naturalist: Theodore Roosevelt, a Lifetime of Exploration, and the Triumph of American Natural History* (New York: Broadway, 2016), John M. MacKenzie's *The Empire of Nature: Hunting, Conservation and British Imperialism* (Manchester, UK: Manchester University Press, 1988), Richard Nelson's *Heart and Blood: Living with Deer in America* (New York: Vintage, 1998), Edward I. Steinhart's *Black Poachers, White Hunters: A Social History of Hunting in Colonial Africa* (Oxford, UK: James Currey, 2006), and Dan Wylie's *Death and Compassion: The Elephant in Southern African Literature* (Johannesburg: Wits University Press, 2018).

In the recent decades the number of important works on the history of animals in captivity (in zoos, circuses, and other settings) and the history of animals in human culture more generally has increased dramatically. On the history of animals in captivity, works especially important to me have been Eric Baratay and Elisabeth Hardouin-Fugier's *Zoo: A History of Zoological Gardens in the West* (London: Reaktion, 2004), Daniel Bender's *The Animal Game: Searching for Wildness at the American Zoo* (Cambridge, MA: Harvard University Press, 2016), Janet M. Davis's *The Circus Age: Culture and Society under the American Big Top* (Chapel Hill: University of North Carolina Press, 2002), Jesse Donahue and Erik Trump's *The Politics of Zoos: Exotic Animals and Their Protectors* (DeKalb: Northern Illinois University Press, 2006), Andrew Flack's, *The Wild Within: Histories of a Landmark British Zoo* (Charlottesville: University of Virginia Press, 2018), David Hancocks's, *A Different Nature: The Paradoxical World of Zoos and Their Uncertain Future* (Berkeley: University of California Press, 2001), Elizabeth Hanson's *Animal Attractions: Nature on Display in American Zoos* (Princeton, NJ: Princeton University Press, 2002), Randy Malamud's, *Reading Zoos: Representations of Animals and Captivity* (New York: New York University Press, 1998), Ian J. Miller's *The Nature of the Beasts: Empire and Exhibition at the Tokyo Imperial Zoo* (Berkeley: University of California Press, 2013), Bob Mullan and Garry Marvin's *Zoo Culture* (London: Weidenfeld and Nicolson, 1987), Susan Nance's, *Entertaining Elephants: Animal Agency and the Business of the American Circus* (Baltimore, MD: Johns Hopkins University Press, 2013), and Lisa Uddin's *Zoo Renewal: White Flight and the Animal Ghetto* (Minneapolis: University of Minnesota Press, 2015).

On animals and human culture, I recommend from a growing field, Steve Baker's *Picturing the Beast: Animals, Identity, and Representation* (Manchester, UK: Manchester University Press, 1993), Mark V. Barrow's *Nature's Ghosts: Confronting Extinction from the Age of Jefferson to the Age of Ecology* (Chicago: University of Chicago Press, 2009), Marcus Baynes-Rock's *Crocodile Undone: The Domestication of Australia's Fauna* (State College: Penn State University Press, 2020), Jonathan Burt's *Animals in Film* (London: Reaktion, 2002), Cynthia Chris's *Watching Wildlife* (Minneapolis: University of Minnesota Press, 2006), Jason M. Colby's *Orca: How We Came to Know and Love the Ocean's Greatest Predator* (New York: Oxford University Press, 2018), Erica Fudge's *Brutal Reasoning: Animals, Rationality, and Humanity in*

Early Modern England (Ithaca, NY: Cornell University Press, 2006), Hal Herzog's *Some We Love, Some We Hate, Some We Eat: Why It's So Hard to Think Straight about Animals* (New York: Harper Perennial, 2010), Akira Mizuta Lippit's *Electric Animal: Toward a Rhetoric of Wildlife* (Minneapolis: University of Minnesota Press, 2008), Susan McHugh's *Animal Stories: Narrating across Species Lines* (Minneapolis: University of Minnesota Press, 2011), Gregg A. Mitman's *Reel Nature: America's Romance with Wildlife on Film* (Cambridge, MA: Harvard University Press, 1999), Lynn K. Nyhart's *Modern Nature: The Rise of the Biological Perspective in Germany* (Chicago: University of Chicago Press, 2009), Rachel Poliquin's *The Breathless Zoo: Taxidermy and the Cultures of Longing* (State College: Penn State University Press, 2012), Harriet Ritvo's *The Animal Estate: The English and Other Creatures in Victorian England* (Cambridge, MA: Harvard University Press, 1987), Louise E. Robbins's *Elephant Slaves and Pampered Parrots: Exotic Animals in Eighteenth-Century Paris* (Baltimore, MD: Johns Hopkins University Press, 2002), Peter Sahlins's *1668: The Year of the Animal in France* (Cambridge, MA: MIT Press, 2017), Nicole Shukin's *Animal Capital: Rendering Life in Biopolitical Times* (Minneapolis: University of Minnesota Press, 2009), Keith Thomas's, *Man and the Natural World: A History of the Modern Sensibility* (New York: Pantheon, 1983), Brett L. Walker's *The Lost Wolves of Japan* (Seattle: University of Washington Press, 2008), and Christian C. Young's *In the Absence of Predators: Conservation and Controversy on the Kaibab Plateau* (Lincoln: University of Nebraska Press, 2002).

Please consult the endnotes for more resources.

INDEX

Page references to photographs and illustrations are in *italic* type.